D0965727

TEACH YOURSELF

# TRIGONOMETRY

NTC Publishing Group

TEACH YOURSELF BOOKS

# TRIGONOMETRY

P. Abbott
and
M. E. Wardle

Long-renowned as *the* authoritative source for self-guided learning – with more than 30 million copies sold worldwide – the *Teach Yourself* series includes over 200 titles in the fields of languages, crafts, hobbies, sports, and other leisure activities.

This edition was first published in 1992 by NTC Publishing Group, 4255 West Touhy Avenue, Lincolnwood (Chicago), Illinois 60646 – 1975 U.S.A. Originally published by Hodder and Stoughton Ltd.


Library of Congress Catalog Card Number: 92-80880

Printed in England by Clays Ltd, St Ives plc.

# Contents

# Introduction

Two major difficulties present themselves when a book of this kind is planned.

In the first place those who use it may desire to apply it in a variety of ways and will be concerned with widely different problems to which trigonometry supplies the solution.

In the second instance the previous mathematical training of its readers will vary considerably.

To the first of these difficulties there can be but one solution. The book can do no more than include those parts which are fundamental and common to the needs of all who require trigonometry to solve their problems. To attempt to deal with the technical applications of the subject in so many different directions would be impossible within the limits of a small volume. Moreover, students of all kinds would find the book overloaded by the inclusion of matter which, while useful to some, would be unwanted by others.

Where it has been possible and desirable, the bearing of certain sections of the subject upon technical problems has been indicated, but, in general, the book aims at putting the student in a position to apply to individual problems the principles, rules and formulae which form the necessary basis for practical applications.

The second difficulty has been to decide what preliminary mathematics should be included in the volume so that it may be intelligible to those students whose previous mathematical equipment is slight. The general aim of the volumes in the series is that, as far as possible, they shall be self-contained. But in this volume it is obviously necessary to assume some previous mathematical training. The study of trigonometry cannot be begun without a knowledge of arithmetic, a certain amount of algebra, and some acquaintance with the fundamentals of geometry.

It may safely be assumed that all who use this book will have a sufficient knowledge of arithmetic. In algebra the student is expected to have studied at least as much as is contained in the volume in this series called *Teach Yourself Algebra*.

The use of an electronic calculator is essential and there can be no progress in the application of trigonometry without having access to a calculating aid. Accordingly chapter 2 is devoted to using a calculator and unless you are reasonably proficient you should not proceed with the rest of the book until you have covered this work. Ideally a scientific calculator is required, but since trigonometric tables are included at the end of the book, it is in fact possible to cover the work using a simple four rule calculator.

No explanation of graphs has been attempted in this volume. In these days, however, when graphical illustrations enter so generally into our daily life, there can be few who are without some knowledge of them, even if no study has been made of the underlying mathematical principles. But, although graphs of trigonometrical functions are included, they are not essential in general to a working knowledge of the subject.

A certain amount of geometrical knowledge is necessary as a foundation for the study of trigonometry, and possibly many who use this book will have no previous acquaintance with geometry. For them chapter 1 has been included. This chapter is in no sense a course of geometry, or of geometrical reasoning, but merely a brief descriptive account of geometrical terms and of certain fundamental geometrical theorems which will make the succeeding chapters more easily understood. It is not suggested that a great deal of time should be spent on this part of the book, and no exercises are included. It is desirable, however, that you make yourself well acquainted with the subject-matter of it, so that you are thoroughly familiar with the meanings of the terms employed and acquire something of a working knowledge of the geometrical theorems which are stated.

The real study of trigonometry begins with chapter 3, and from that point until the end of chapter 9 there is very little that can be omitted by any student. Perhaps the only exception is the 'product formulae' in sections 86–88. This section is necessary, however, for the proof of the important formula of section 98, but a student who is pressed for time and finds this part of the work troublesome, may be content to assume the truth of it when studying section 98. In chapter 9 you will reach what you may consider the goal of elementary trigonometry, the 'solution of the triangle' and its many applications, and there you may be content to stop.

Chapters 10, 11 and 12 are not essential for all practical applications of the subject, but some students, such as electrical engineers and, of course, all who intend to proceed to more advanced work, cannot afford to omit them. It may be noted that previous to chapter 9 only angles which are not greater than 180° have been considered, and these have been taken in two stages in chapters 3 and 5, so that the approach may be easier. Chapter 11 continues the work of these two chapters and generalises with a treatment of angles of any magnitude.

The exercises throughout have been carefully graded and selected in such a way as to provide the necessary amount of manipulation. Most of them are straightforward and purposeful; examples of academic interest or requiring special skill in manipulation have, generally speaking, been excluded.

Trigonometry employs a comparatively large number of formulae. The more important of these have been collected and printed on pp. 171–173 in a convenient form for easy reference.

# 1

# Geometrical Foundations

## 1 Trigonometry and Geometry

The name trigonometry is derived from the Greek words meaning 'triangle' and 'to measure'. It was so called because in its beginnings it was mainly concerned with the problem of 'solving a triangle'. By this is meant the problem of finding all the sides and angles of a triangle, when some of these are known.

Before beginning the study of trigonometry it is desirable, in order to reach an intelligent understanding of it, to acquire some knowledge of the fundamental geometrical ideas upon which the subject is built. Indeed, geometry itself is thought to have had its origins in practical problems which are now solved by trigonometry. This is indicated in certain fragments of Egyptian mathematics which are available for our study. We learn from them that, from early times, Egyptian mathematicians were concerned with the solution of problems arising out of certain geographical phenomena peculiar to that country. Every year the Nile floods destroyed landmarks and boundaries of property. To re-establish them, methods of surveying were developed, and these were dependent upon principles which came to be studied under the name of 'geometry'. The word 'geometry', a Greek one, means 'Earth measurement', and this serves as an indication of the origins of the subject.

We shall therefore begin by a brief consideration of certain geometrical principles and theorems, the applications of which we shall subsequently employ. It will not be possible, however, within this small book to attempt mathematical proofs of the various theorems which will be stated. The student who has not previously approached the subject of geometry, and who desires to acquire a more complete knowledge of it, should turn to any good modern treatise on this branch of mathematics.

## 2 The Nature of Geometry

Geometry has been called 'the science of space'. It deals with solids, their forms and sizes. By a 'solid' we mean a *portion of space bounded by surfaces*, and in geometry we deal only with what are called *regular solids*. As a simple example consider that familiar solid, the cube. We are not concerned with the material of which it is composed, but merely the shape of the portion of space which it occupies. We note that it is bounded by six surfaces, which are squares. Each square is said to be at right angles to adjoining squares. Where two squares intersect **straight lines** are formed; three adjoining squares meet in a **point**. These are examples of some of the matters that geometry considers in connection with this particular solid.

For the purpose of examining the geometrical properties of the solid we employ a conventional representation of the cube, such as is shown in Fig. 1. In this, all the faces are shown, as though the body were made of transparent material, those edges which could not otherwise be seen being indicated by dotted lines. The student can follow from this figure the properties mentioned above.

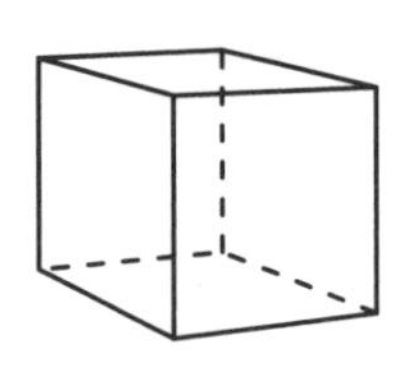

Fig. 1.

## 3 Plane surfaces

The surfaces which form the boundaries of the cube are level or flat surfaces, or in geometrical terms **plane surfaces**. It is important that the student should have a clear idea of what is meant by a plane surface. It may be described as a *level* surface, a term that everybody understands although they may be unable to give a mathematical definition of it. Perhaps the best example in nature of a level surface or plane surface is that of still water. A water surface is also a **horizontal surface**.

The following definition will present no difficulty to the student.

*A plane surface is such that the straight line which joins any two points on it lies wholly in the surface.*

It should further be noted that

A plane surface is determined uniquely, by

(*a*) Three points not in the same straight line,
(*b*) Two intersecting straight lines.

By this we mean that one plane, and one only, can include (*a*) three given points, or (*b*) two given intersecting straight lines.

It will be observed that we have spoken of surfaces, points and straight lines without defining them. Every student probably understands what the terms mean, and we shall not consider them further here, but those who desire more precise knowledge of them should consult a geometrical treatise. We shall now consider theorems connected with points and lines on a plane surface. This is the part of geometry called *plane geometry*. The study of the shapes and geometrical properties of solids is the function of *solid geometry*, which we will touch on later.

## 4 Angles

Angles are of the utmost importance in trigonometry, and the student must therefore have a clear understanding of them from the outset. Everybody knows that an angle is formed when two straight lines of two surfaces meet. This has been assumed in section 2. But a precise mathematical definition is helpful. Before proceeding to that, however, we will consider some elementary notions and terms connected with an angle.

Fig. 2(a), (b), (c) show three examples of angles.

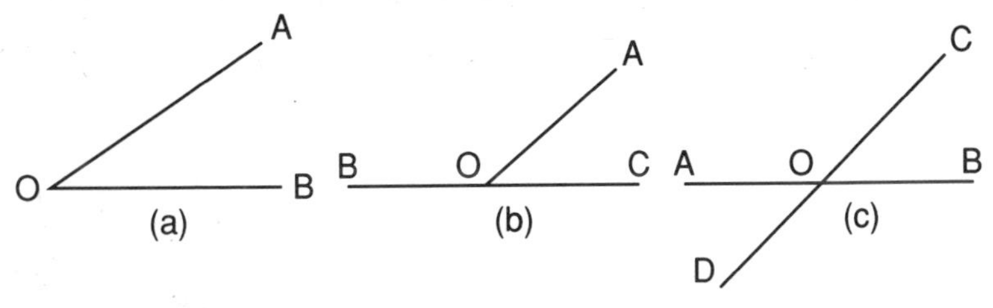


**Fig. 2.**

(1) In Fig. 2(a) two straight lines OA, OB, called the **arms** of the angle, meet at O to form the angle denoted by AOB.

O is termed the **vertex** of the angle.

The arms may be of any length, and the size of the angle is not altered by increasing or decreasing them.

The 'angle AOB' can be denoted by ∠AOB or AÔB. It should be noted that the middle letter, in this case O, always indicates the vertex of the angle.

(2) In Fig. 2(b) the straight line AO is said to **meet** the straight line CB at O. Two angles are formed, AOB and AOC, with a common vertex O.

(3) In Fig. 2(c) two straight lines AB and CD cut one another at O. Thus there are formed four angles COB, AOC, DOA, DOB.

The pair of angles COB, AOD are termed **vertically opposite** angles. The angles AOC, BOD are also vertically opposite.

### Adjacent angles

Angles which have a common vertex and also one common arm are called **adjacent angles**. Thus in Fig. 2(b) AOB, AOC are adjacent, in Fig. 2(c) COB, BOD are adjacent, etc.

## 5 Angles formed by rotation

We must now consider a mathematical conception of an angle.

Imagine a straight line, starting from a fixed position on OA (Fig. 3) rotating about a point O in the direction indicated by an arrow.

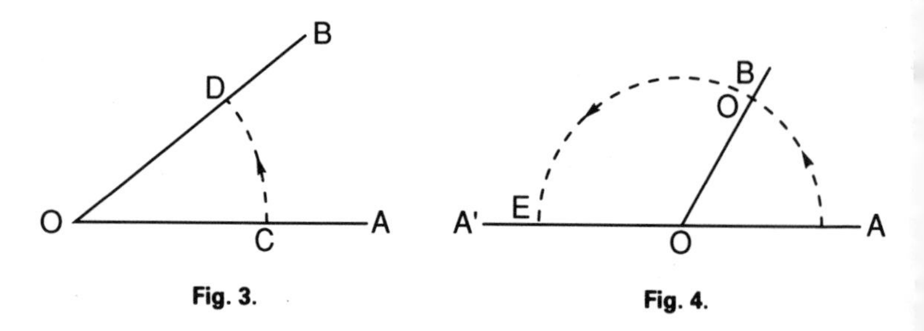


Fig. 3. Fig. 4.

Let it take up the position indicated by OB.

In rotating from OA to OB an angle AOB is **described**.

Thus we have the conception of an angle as formed by the rotation of a straight line about a fixed point, the vertex of the angle.

If any point C be taken on the rotating arm, it will clearly mark out an arc of a circle, CD.

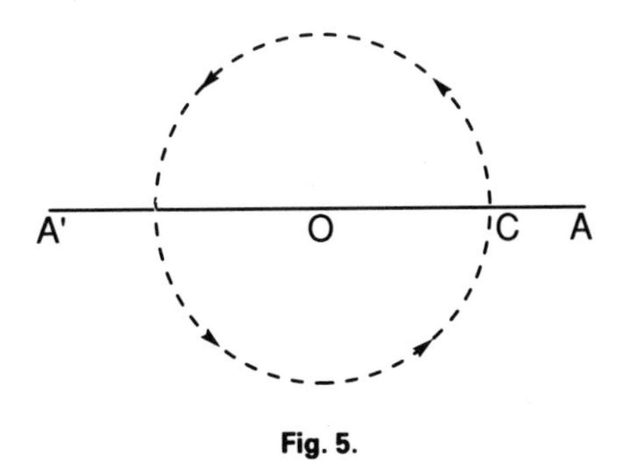


**Fig. 5.**

There is no limit to the amount of rotation of OA, and consequently angles of any size can be formed by a straight line rotating in this way.

### A half rotation

Let us next suppose that the rotation from OA to OB is continued until the position OA′ is reached (Fig. 4), in which OA′ and OA are in the same straight line. The point C will have marked out a semi-circle and the angle formed AOA′ is sometimes called a **straight angle**.

### A complete rotation

Now let the rotating arm continue to rotate, in the same direction as before, until it arrives back at its original position on OA. It has then made a **complete rotation**. The point C, on the rotating arm, will have marked out the circumference of a circle, as indicated by the dotted line.

## 6 Measurement of angles

### (a) Sexagesimal measure

The conception of formation of an angle by rotation leads us to a convenient method of measuring angles. We imagine the complete rotation to be divided into 360 equal divisions; thus we get 360 small equal angles, each of these is called a degree, and is denoted by **1°**.

Since any point on the rotating arm marks out the circumference of a circle, there will be 360 equal divisions of this circumference,

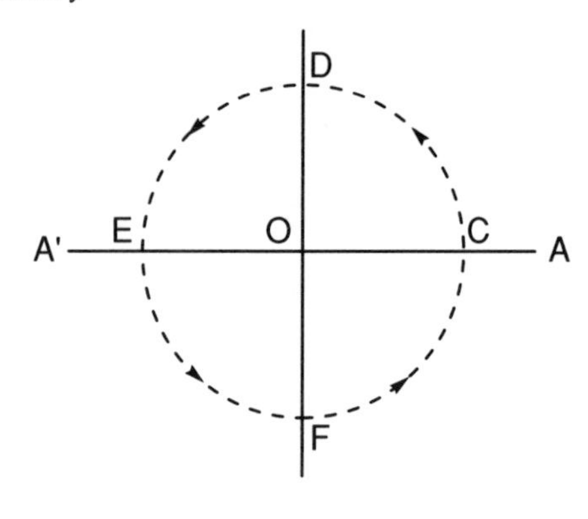


**Fig. 6.**

corresponding to the 360 degrees (see Theorem 17). If these divisions are marked on the circumference we could, by joining the points of division to the centre, show the 360 equal angles. These could be numbered, and thus the figure could be used for measuring any given angle. In practice the divisions and the angles are very small, and it would be difficult to draw them accurately. This, however, is the principle of the **circular protractor**, which is an instrument devised for the purpose of measuring angles. Every student of trigonometry should be equipped with a protractor for this purpose.

### *Right angles*

Fig. 6 represents a complete rotation, such as was shown in Fig. 5. Let the points D and F be taken half-way between C and E in each semi-circle.

The circumference is thus divided into four equal parts.

The straight line DF will pass through O.

The angles COD, DOE, EOF, FOC, each one quarter of a complete rotation, are termed **right angles**, and each contains 90°.

The circle is divided into four equal parts called **quadrants**, and numbered the first, second, third and fourth quadrants in the order of their formation.

Also when the rotating line has made a half rotation, the angle formed – the straight angle – must contain 180°.

Each degree is divided into 60 minutes, shown by ′.

Each minute is divided into 60 seconds, shown by ″.

Thus 37° 15′ 27″ means an angle of

37 degrees, 15 minutes, 27 seconds.
or 37.2575 degrees, correct to 4 decimal places

*Note*, $30' = 0.5°$, $1' = 1/60° = 0.01667°$,
$1'' = 1/(60 \times 60)° = 0.0002778°$

This division into so many small parts is very important in navigation, surveying, gunnery, etc., where great accuracy is essential.

For the purpose of this book we shall give results correct to the nearest 1/100th of a degree, i.e. correct to 2 decimal places.

*Historical note.* The student may wonder why the number 360 has been chosen for the division of a complete rotation to obtain the degree. The selection of this number was made in very early days in the history of the world, and we know, for example, from inscriptions that it was employed in ancient Babylon. The number probably arose from the division of the heavens by ancient astronomers into 360 parts, corresponding to what was reputed to be the number of days in the year. The number 60 was possibly used as having a large number of factors and so capable of being used for easy fractions.

### (b) Centesimal measure

When the French adopted the metric system they abandoned the method of dividing the circle into 360 parts. To make the system of measuring angles consistent with other metric measures, it was decided to divide the right angle into 100 equal parts, and consequently the whole circle into 400 parts. The angles thus obtained were called **grades**.

Consequently 1 right angle = 100 grades.
1 grade = 100 minutes.
1 minute = 100 seconds.

### (c) Circular measure

There is a third method of measuring angles which is an **absolute** one, that is, it does not depend upon dividing the right angle into any arbitrary number of equal parts, such as 360 or 400.

The unit is obtained as follows:

In a circle, centre O (see Fig. 7), let a radius OA rotate to a position OB, such that the length of the **arc** AB is equal to that of the radius.

In doing this an angle AOB is formed which is the unit of measurement. It is called a **radian**. The size of this angle will be the same whatever radius is taken. It is absolute in magnitude.

In degrees *1 radian = 57° 17′ 44.8″* (approx.) or 57.29578°. This method of measuring angles will be dealt with more fully in chapter 10. It is very important and is always used in the higher branches of mathematics.

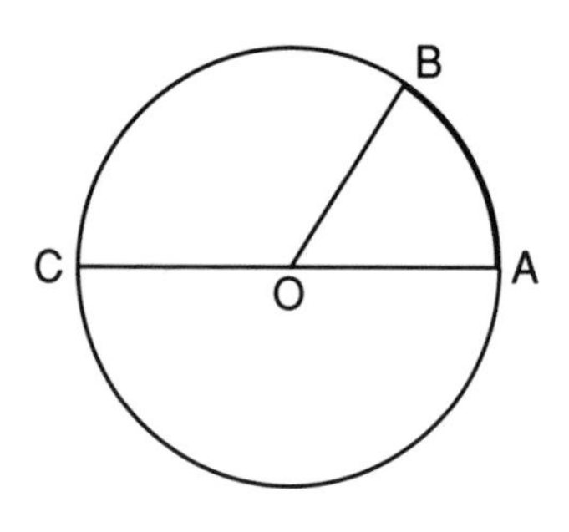


**Fig. 7.**

## 7 Terms used to decribe angles

An **acute angle** is an angle which is *less* than a right angle.

An **obtuse angle** is one which is *greater* than a right angle.

**Reflex** or **re-entrant angles** are angles between 180° and 360°.

**Complementary angles.** When the sum of two angles is equal to a right angle, each is called the *complement* of the other. Thus the complement of 38° is 90° − 38° = 52°.

**Supplementary angles.** When the sum of two angles is equal to 180°, each angle is called the *supplement* of the other. Thus the supplement of 38° is 180° − 38° = 142°.

## 8 Geometrical Theorems

We will now state, without proof, some of the more important geometrical theorems.

### Theorem 1 Intersecting straight lines

*If two straight lines intersect, the vertically opposite angles are equal.* (See section 4.)

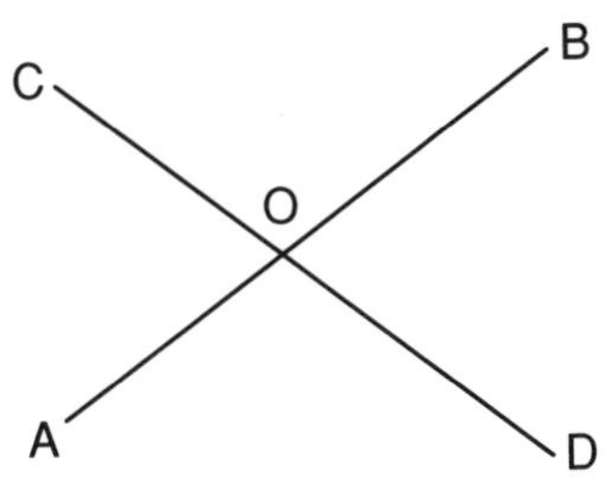


**Fig. 8.**

In Fig. 8, AB and CD are two straight lines intersecting at O.

Then $\angle AOC = \angle BOD$

and $\angle COB = \angle AOD$

The student will probably see the truth of this on noticing that $\angle AOC$ and $\angle BOD$ are each supplementary to the same angle, COB.

## 9 Parallel straight lines

Take a set square PRQ (Fig. 9) and slide it along the edge of a ruler.

Let $P_1R_1Q_1$ be a second position which it takes up.

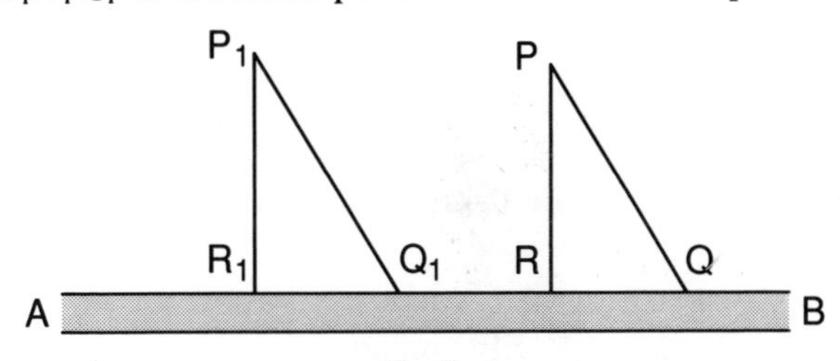


**Fig. 9.**

It is evident that the inclination of PQ to AB is the same as that of $P_1Q_1$ to AB, since there has been no change in direction.

$$\therefore \angle PQB = \angle P_1Q_1B$$

If PQ and $P_1Q_1$ were produced to any distance they would not meet.

The straight lines PQ and $P_1Q_1$ are said to be **parallel**.

Similarly PR and $P_1R_1$ are parallel.

Hence the following definition.

*Straight lines in the same plane which will not meet however far they may be produced are said to be parallel.*

## Direction

*Parallel straight lines in a plane have the same direction.*

If a number of ships, all sailing north in a convoy are ordered to change direction by turning through the same angle they will then follow parallel courses.

## Terms connected with parallel lines

In Fig. 10 AB, CD represent two parallel straight lines.

***Transversal***

A straight line such as PQ which cuts them is called a **transversal**.

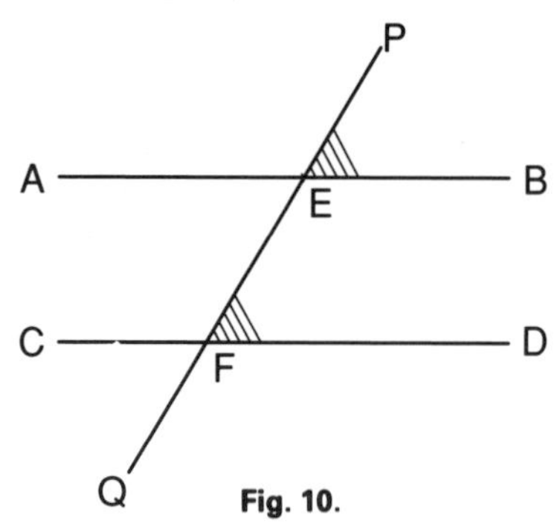


**Fig. 10.**

***Corresponding angles***

On each side of the transversal are two pairs of angles, one pair of which is shaded in the figure. These are called **corresponding angles**.

***Alternate angles***

Two angles such as AEF, EFD on opposite sides of the transversal are called **alternate angles**.

## Theorem 2

*If a pair of parallel straight lines be cut by a transversal*

(*a*) *alternate angles are equal,*
(*b*) *corresponding angles on the same side of the transversal are equal,*
(*c*) *the two interior angles on the same side of the transversal are equal to two right angles.*

Thus in Fig. 10:
Alternate angles: $\angle AEF = \angle EFD$; $\angle BEF = \angle EFC$.

Corresponding angles: $\angle PEB = \angle EFD$; $\angle BEF = \angle DFQ$.
Similarly on the other side of the transversal:
Interior angles: $\angle AEF + \angle EFD = 2$ right angles,
also $\angle AEF + \angle EFC = 2$ right angles.

## 10 Triangles

### Kinds of triangles

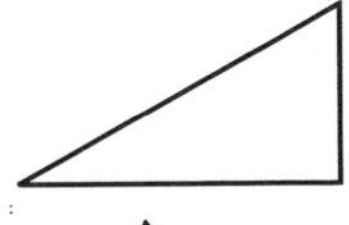

A **right-angled triangle** has one of its angles a right angle. The side opposite to the right angle is called the **hypotenuse**.

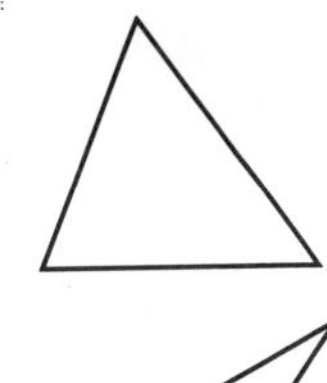

An **acute-angled triangle** has all its angles acute angles (see section 7).

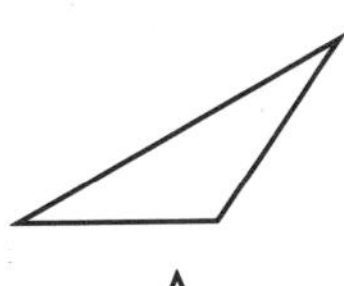

An **obtuse-angled triangle** has one of its angles obtuse (see section 7).

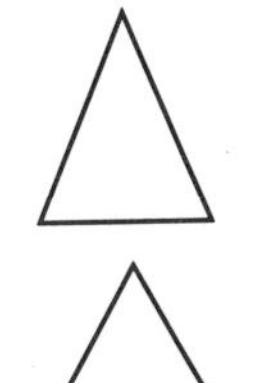

An **isosceles triangle** has two of its sides equal.

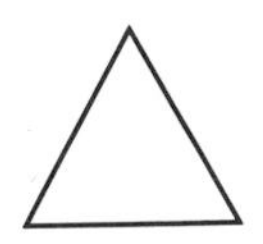

An **equilateral triangle** has all its sides equal.

**Fig. 11.**

### Lines connected with a triangle

The following terms are used for certain lines connected with a triangle.

In $\triangle$ ABC, Fig. 12,

(1) AP is the perpendicular from A to BC. It is called the **altitude** from the vertex A.
(2) AQ is the **bisector of the vertical angle** at A.

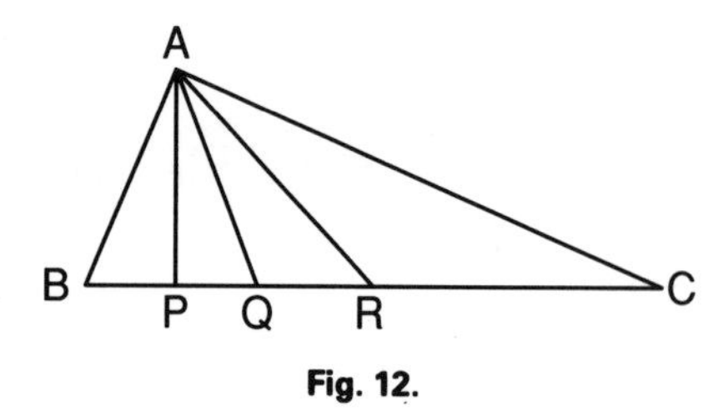


**Fig. 12.**

(3) AR **bisects** BC. It is called a **median**. If each of the points B and C be taken as a vertex, there are two other corresponding medians. Thus a triangle may have three medians.

## 11 Theorem 3 Isosceles and equilateral triangles

*In an isosceles triangle*

(*a*) *The sides opposite to the equal angles are equal,*
(*b*) *A straight line drawn from the vertex perpendicular to the opposite side bisects that side and the vertical angle.*

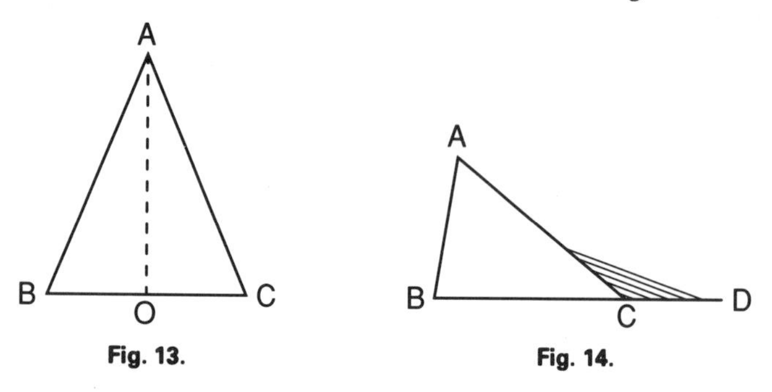


**Fig. 13.** **Fig. 14.**

In Fig. 13, ABC is an isosceles △ and AO is drawn perpendicular to the base from the vertex A.

Then by the above $\angle ABC = \angle ACB$
$BO = OC$
$\angle BAO = \angle CAO.$

### Equilateral triangle

The above is true for an equilateral triangle, and since all its sides are equal, all its angles are equal.

*Note*, in an isosceles △ the altitude, median and bisector of the vertical angle (see section 10) coincide when the point of intersection of the two equal sides is the vertex. If the △ is equilateral they coincide for all three vertices.

## 12 Angle properties of a triangle

### Theorem 4

*If one side of a triangle be produced, the exterior angle so formed is equal to the sum of the two interior opposite angles.*

Thus in Fig. 14 one side BC of the △ ABC is produced to D.

∠ACD is called an **exterior** angle.

Then by the above

$$\angle ACD = \angle ABC + \angle BAC$$

*Notes*

(1) Since the exterior angle is equal to the sum of the opposite interior angles, it must be greater than either of them.
(2) As each side of the triangle may be produced in turn, there are three exterior angles.

### Theorem 5

*The sum of the angles of any triangle is equal to two right angles.*

*Notes* It follows that:

(1) each of the angles of an equilateral triangle is 60°,
(2) in a right-angled triangle the two acute angles are complementary (see section 7),
(3) the sum of the angles of a quadrilateral is 360° since it can be divided into two triangles by joining two opposite points.

## 13 Congruency of triangles

*Triangles which are equal in all respects are said to be congruent.*

Such triangles have corresponding sides and angles equal, and are exact copies of one another.

If two triangles ABC and DEF are congruent we may express this by the notation △ ABC ≡ △ DEF.

### Conditions of congruency

Two triangles are congruent when

### Theorem 6

*Three sides of one are respectively equal to the three sides of the other.*

## Theorem 7

*Two sides of one and the angle they contain are equal to two sides and the contained angle of the other.*

## Theorem 8

*Two angles and a side of one are equal to two angles and the corresponding side of the other.*

These conditions in which triangles are congruent are very important. The student can test the truth of them practically by constructing triangles which fulfil the conditions stated above.

## The ambiguous case

The case of constructing a triangle when two sides and an angle opposite to one of them are given, not contained by them as in Theorem 7, requires special consideration.

*Example.* Construct a triangle in which two sides are 35 mm and 25 mm and the angle opposite the smaller of these is 30°.

The construction is as follows:

Draw a straight line AX of indefinite length (Fig. 15).
At A construct ∠BAX = 30° and make AB = 35 mm.
With B as centre and radius 25 mm construct an arc of a circle to cut AX.
This it will do in two points, C and C′.
Consequently if we join BC or BC′ we shall complete two triangles ABC, ABC′ each of which will fulfil the given conditions. There being thus two solutions the case is called **ambiguous**.

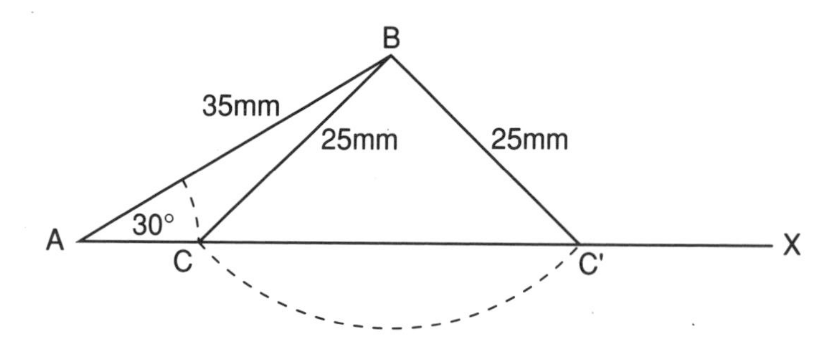


**Fig. 15.**

## 14 Right-angled triangles

### Theorem of Pythagoras (Theorem 9)

*In every right-angled triangle the square on the hypotenuse is equal to the sum of the squares on the sides containing the right angle.*

In Fig. 16 ABC is a right-angled triangle, AB being the hypotenuse. On the three sides squares have been constructed. Then the area of the square described on AB is equal to the sum of the areas of the squares on the other two sides.

This we can write in the form

$$AB^2 = AC^2 + BC^2$$

If we represent the length of AB by c, AC by b and BC by a, then $c^2 = a^2 + b^2$.

It should be noted that by using this result, if any two sides of a right-angled triangle are known, we can find the other side for

$$a^2 = c^2 - b^2$$
$$b^2 = c^2 - a^2.$$

*Note* This theorem is named after **Pythagoras**, the Greek mathematician and philosopher who was born about 569 BC. It is one of the most important and most used of all geometrical theorems.

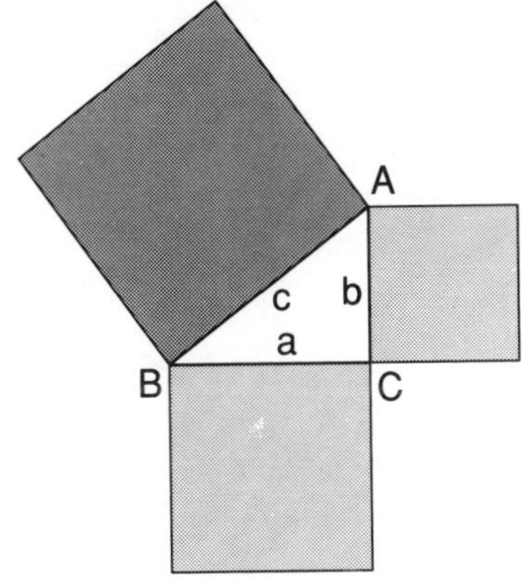


**Fig. 16.**

## 15 Similar triangles

*Definition. If the angles of one triangle are respectively equal to the angles of another triangle the two triangles are said to be similar.*

The sides of similar triangles which are opposite to equal angles in each are called **corresponding sides**.

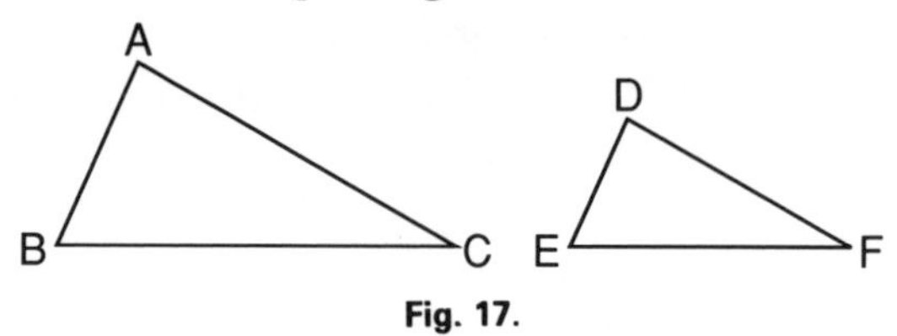


**Fig. 17.**

In Fig. 17 the triangles ABC, DEF are equiangular

$$\angle ABC = \angle DEF,$$
$$\angle BAC = \angle EDF,$$
$$\angle ACB = \angle DFE.$$

The sides AB, DE are two corresponding sides.
So also are AC and DF, BC and EF.
Fig 18 shows another example of interest later.
AB, CD EF are parallel.
Then by the properties of parallel lines (see section 9)

$$\angle OAB = \angle OCD = \angle OEF$$

also $$\angle OBA = \angle ODC = \angle OFE.$$

∴ the triangles OAB, OCD, OEF are similar.

## Property of similar triangles

### Theorem 10

*If two triangles are similar, the corresponding sides are proportional.*

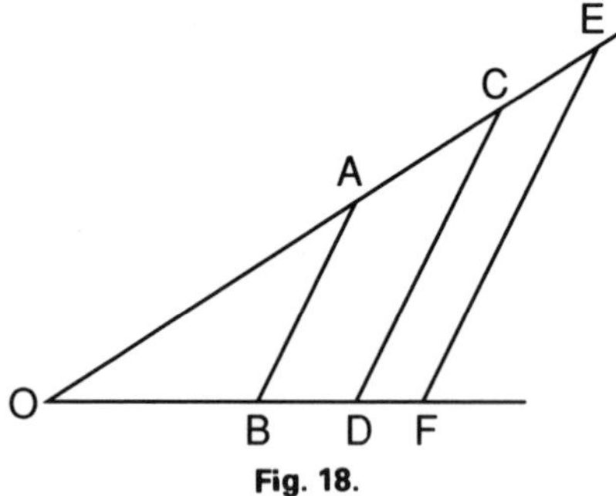


**Fig. 18.**

Thus in Fig. 17:

$$\frac{AB}{BC} = \frac{DE}{EF}, \frac{AB}{AC} = \frac{DE}{DF}, \frac{AC}{CB} = \frac{DF}{FE}$$

Similarly in Fig. 18:

$$\frac{AB}{BO} = \frac{CD}{DO} = \frac{EF}{FO},$$

$$\frac{AB}{OA} = \frac{CD}{OC} = \frac{EF}{OE}, \text{ etc.}$$

These results are of great importance in trigonometry.

*Note* A similar relation holds between the sides of quadrilaterals and other rectilinear figures which are equiangular.

## 16 Quadrilaterals

A **quadrilateral** is a plane figure with four sides, and a straight line joining two opposite angles is called a **diagonal**.

The following are among the principal quadrilaterals, with some of their properties:

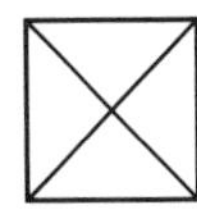
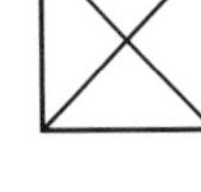

(1) The **square** (a) has all its sides equal and all its angles right angles, (b) its diagonals are equal, bisect each other at right angles and also bisect the opposite angles.

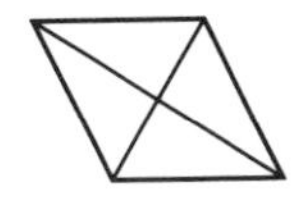

(2) The **rhombus** (a) has all its sides equal, (b) its diagonals bisect each other at right angles and bisect the opposite angles.

(3) The **rectangle** (a) has opposite sides equal and all its angles are right angles, (b) its diagonals are equal and bisect each other.

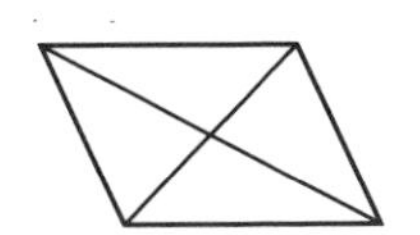

(4) The **parallelogram** (a) has opposite sides equal and parallel, (b) its opposite angles are equal, (c) its diagonals bisect each other.

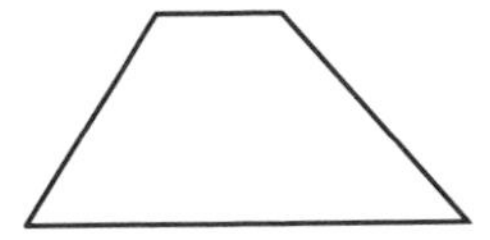

(5) The **trapezium** has two opposite sides parallel.

**Fig. 19.**

## 17 The Circle

It has already been assumed that the student understands what a circle is, but we now give a geometrical definition.

*A circle is a plane figure bounded by one line which is called the circumference and is such that all straight lines drawn to the circumference from a point within the circle, called the centre, are equal.*

These straight lines are called **radii**.

An **arc** is a part of the circumference.

A **chord** is a straight line joining two points on the circumference and dividing the circle into two parts.

A **diameter** is a chord which passes through the centre of the circle. It divides the circle into two equal parts called **semi-circles**.

A **segment** is a part of a circle bounded by a chord and the arc which it cuts off. Thus in Fig. 20 the chord PQ divides the circle into two segments. The larger of these PCQ is called a **major segment** and the smaller, PBQ, is called a **minor segment**.

A **sector** of a circle is that part of the circle which is bounded by two radii and the arc intercepted between them.

Thus in Fig. 21 the figure OPBQ is a sector bounded by the radii OP, OQ and the arc PBQ.

An angle in a segment is the angle formed by joining the ends of a chord or arc to a point on the arc of the segment.

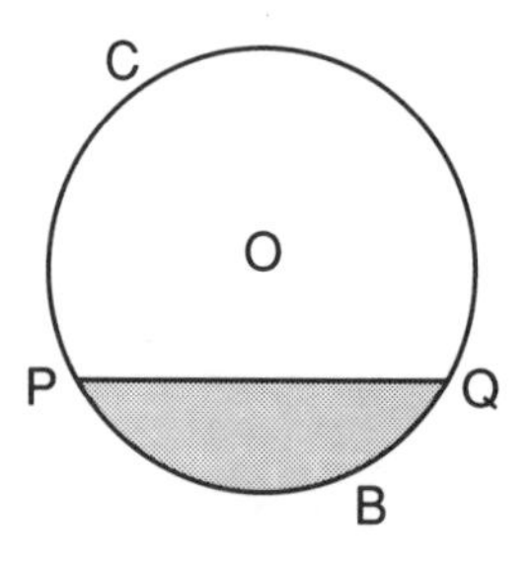

**Fig. 20.**

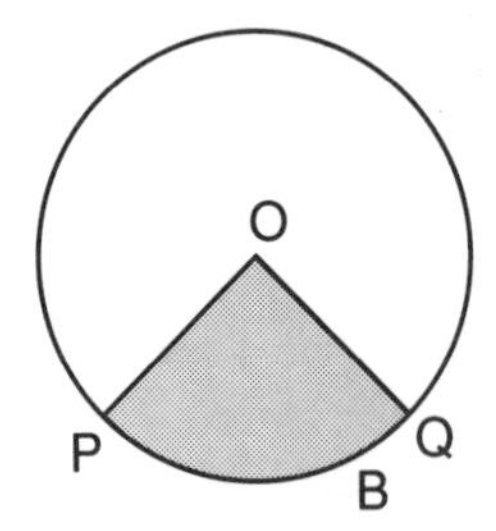

**Fig. 21.**

Thus in Fig. 22, the ends of the chord AB are joined to D a point on the arc of the segment. The angle ADB is the angle in the segment ABCD.

If we join A and B to any point D′ in the minor segment, then ∠AD′B is the angle in the minor segment.

If A and B are joined to the centre O, the angle OB is called the **angle at the centre**.

The angle ADB is also said to **subtend** the arc AB and the ∠AOB is said to be the angle subtended at the centre by the arc AB of the chord AB.

**Concentric Circles** are circles which have the same centre.

## 18 Theorems relating to the circle

### Theorem 11

*If a diameter bisects a chord, which is not a diameter, it is perpendicular to the chord.*

### Theorem 12

*Equal chords in a circle are equidistant from the centre.*

### Theorem 13

*The angle which is subtended at the centre of a circle by an arc is double the angle subtended at the circumference.*

In Fig. 23 ∠AOB is the angle subtended at O the centre of the circle by the arc AB, and ∠ADB is an angle at the circumference (see section 17) as also is ∠ACB.

Then $\angle AOB = 2\angle ADB$

and $\angle AOB = 2\angle ACB$.

### Theorem 14

*Angles in the same segment of a circle are equal to one another.*

In Fig. 23 $\angle ACB = \angle ADB$.

This follows at once from Theorem 13.

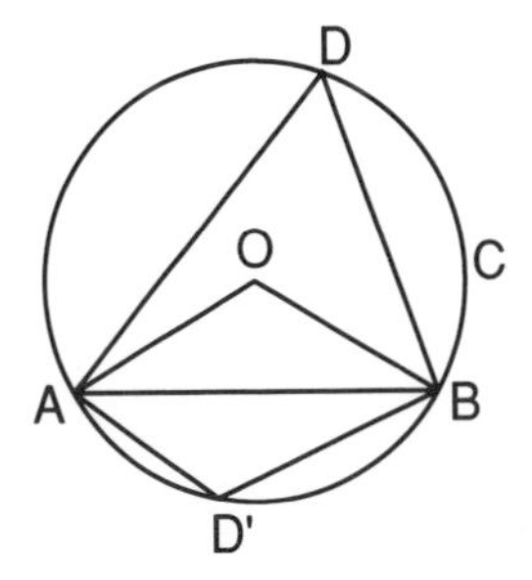


Fig. 22.

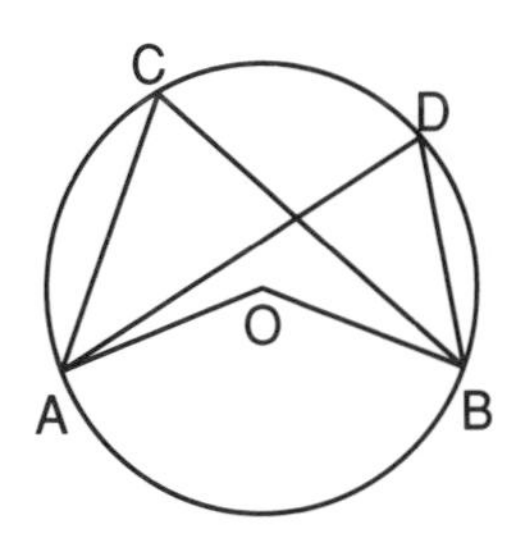


Fig. 23.

### Theorem 15

*The opposite angles of a quadrilateral inscribed in a circle are together equal to two right angles.*

They are therefore supplementary (see section 7).

*Note* A quadrilateral inscribed in a circle is called a **cyclic** or **concyclic** quadrilateral.

In Fig. 24, ABCD is a cyclic quadrilateral.

Then $\angle ABC + \angle ADC = 2$ right angles
$\angle BAD + \angle BCD = 2$ right angles.

### Theorem 16

*The angle in a semi-circle is a right angle.*

In Fig. 25 AOB is a diameter.

The $\angle ACB$ is an angle in one of the semi-circles so formed.

$\angle ACB$ is a right angle.

### Theorem 17

*Angles at the centre of a circle are proportional to the arcs on which they stand.*

In Fig. 26,

$$\frac{\angle POQ}{\angle QOR} = \frac{\text{arc PQ}}{\text{arc QR}}.$$

It follows from this that *equal angles stand on equal arcs.*

This is assumed in the method of measuring angles described in section 6(a).

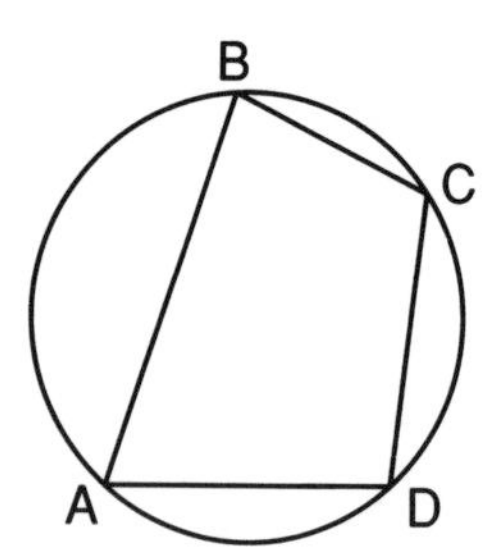


**Fig. 24.**

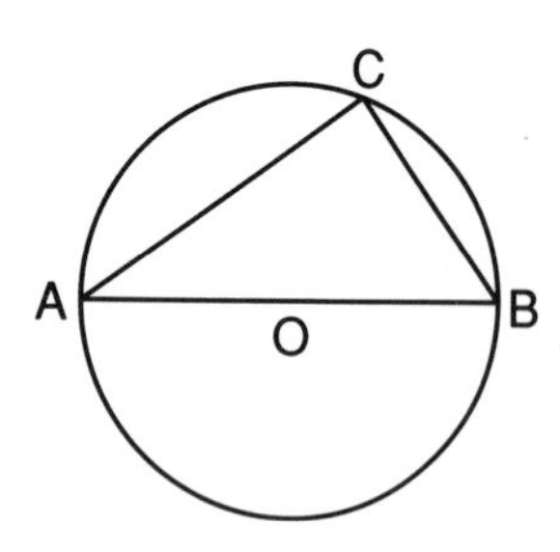


**Fig. 25.**

### Tangent to a circle

*A tangent to a circle is a straight line which meets the circumference of the circle but which when produced does not cut it.*

In Fig. 27 PQ represents a tangent to the circle at a point A on the circumference.

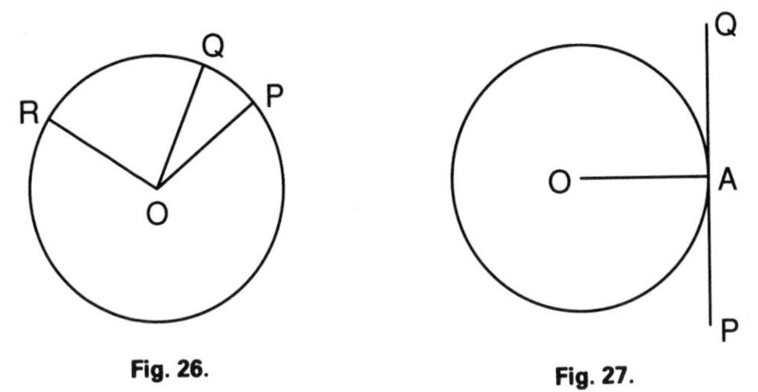


Fig. 26.

Fig. 27.

### Theorem 18

*A tangent to a circle is perpendicular to the radius drawn from the point of contact.*

Thus in Fig. 27 PQ is at right angles to OA.

## 19 Solid geometry

We have so far confined ourselves to the consideration of some of the properties of figures drawn on plane surfaces. In many of the practical applications of geometry we are concerned also with **solids** to which we referred in section 2. In addition to these, in surveying and navigation problems for example, we need to make observations and calculations in different **planes** which are not specifically the surfaces of solids. Examples of these, together with a brief classification of the different kinds of regular solids, will be given later.

## 20 Angle between two planes

Take a piece of fairly stout paper and fold it in two. Let AB, Fig. 28, be the line of the fold. Draw this straight line. Let BCDA, BEFA represent the two parts of the paper.

These can be regarded as two separate planes. Starting with the two parts folded together, keeping one part fixed the other part

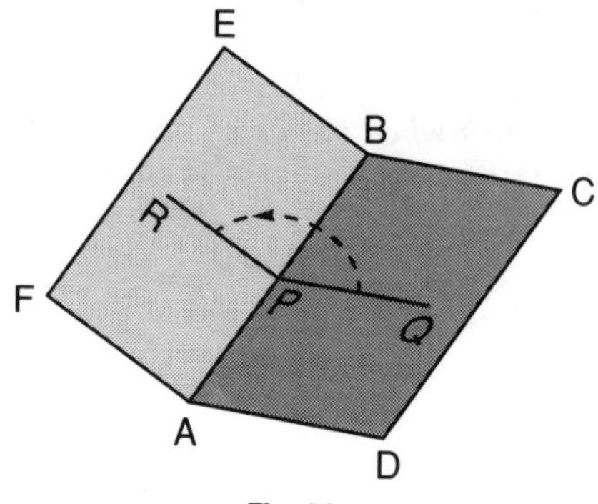


Fig. 28.

can be rotated about AB into the position indicated by ABEF. In this process the one plane has moved through an angle relative to the fixed plane. This is analogous to that of the rotation of a line as described in section 5. We must now consider how this angle can be definitely fixed and measured. Flattening out the whole paper again take any point P on the line of the fold, i.e. AB, and draw RPQ at right angles to AB. If you fold again PR will coincide with PQ. Now rotate again and the line PR will mark out an angle relative to PQ as we saw in section 5. The angle RPQ is thus the angle which measures the amount of rotation, and is called the angle between the planes.

*Definition. The angle between two planes is the angle between two straight lines which are drawn, one in each plane, at right angles to the line of intersection of the plane and from the same point on it.*

When this angle becomes a right angle the planes are **perpendicular** to one another.

As a particular case a plane which is perpendicular to a horizontal plane is called a **vertical plane** (see section 3).

If you examine a corner of the cube shown in Fig. 1 you will see that it is formed by three planes at right angles to one another. A similar instance may be observed in the corner of a room which is rectangular in shape.

## 21 A straight line perpendicular to a plane

Take a piece of cardboard AB (Fig. 29), and on it draw a number of straight lines intersecting at a point O. At O fix a pin OP so that it is perpendicular to all of these lines. Then OP is said to be perpendicular to the plane AB.

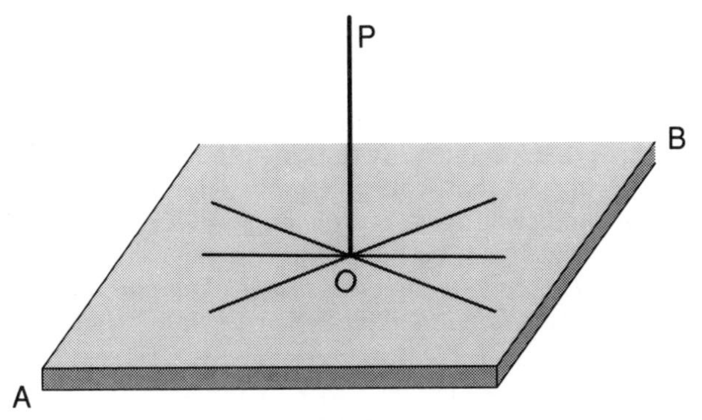


**Fig. 29.**

*Definition. A straight line is said to be perpendicular to a plane when it is perpendicular to any straight line which it meets in the plane.*

### Plumb line and vertical

Builders use what is called a plumb line to obtain a vertical line. It consists of a small weight fixed to a fine line. This vertical line is perpendicular to a horizontal plane.

## 22 Angle between a straight line and a plane

Take a piece of cardboard ABCD, Fig. 30, and at a point O in it fix a needle ON at any angle. At any point P on the needle stick another needle PQ into the board, and perpendicular to it.

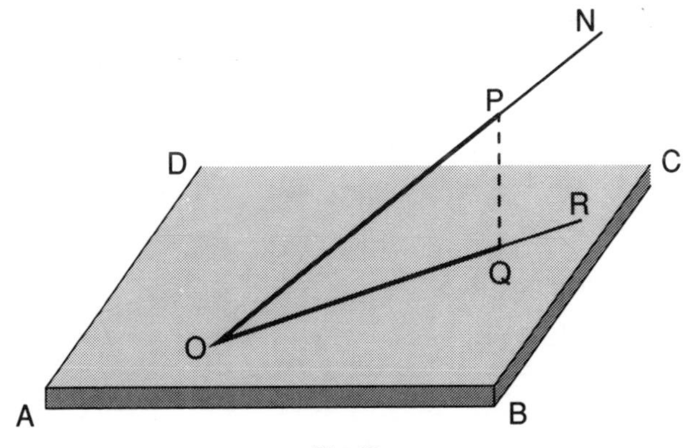


**Fig. 30.**

Draw the line OQR on the board.

OQ is called the **projection** of OP on the plane ABCD.

The angle POQ between OP and its projection on the plane is called the angle between OP and the plane.

If you were to experiment by drawing other lines from O on the plane you will see that you will get angles of different sizes between ON and such lines. But the angle POQ is the smallest of all the angles which can be formed in this way.

*Definition. The angle between a straight line and a plane is the angle between the straight line and its projection on the plane.*

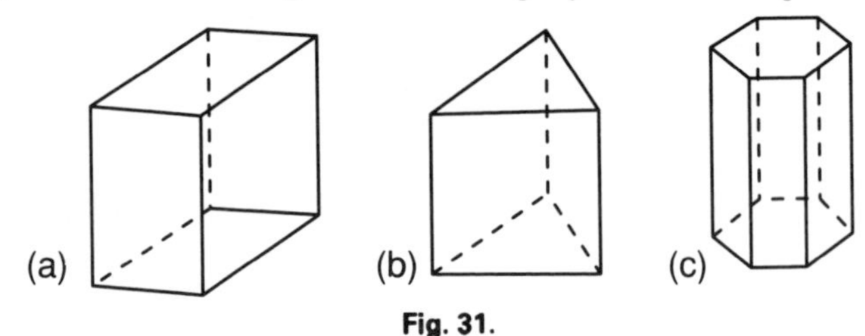


Fig. 31.

## 23 Some regular solids

### (1) Prisms

In Fig. 31 (a), (b), (c) are shown three typical prisms.

(a) is rectangular, (b) is triangular and (c) is hexagonal.

They have two identically equal ends or bases and a rectangle, triangle and regular hexagon respectively.

The sides are rectangles in all three figures and their planes are perpendicular to the bases.

Such prisms are called **right prisms**.

If sections are made parallel to the bases, all such sections are identically equal to the bases. A prism is a solid with a **uniform cross section**.

Similarly other prisms can be constructed with other geometrical figures as bases.

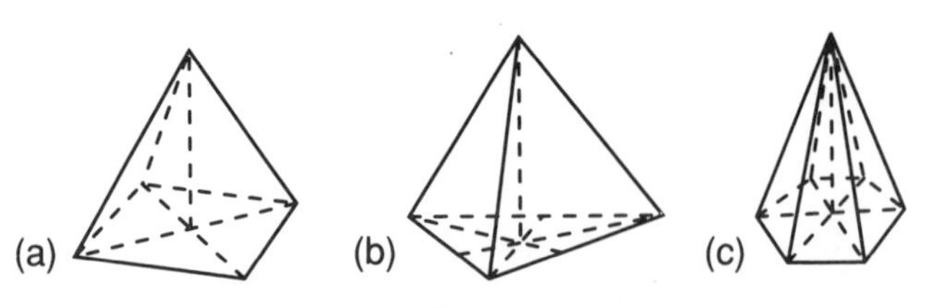


Fig. 32.

## (2) Pyramids

In Fig. 32 (a), (b), (c), are shown three typical pyramids.

(a) is a square pyramid, (b) is a triangular pyramid, (c) is a hexagonal pyramid.

Pyramids have one base only, which, as was the case with prisms, is some geometrical figure.

The sides, however, are isosceles triangles, and they meet at a point called the **vertex**.

The angle between each side and the base can be determined as follows for a square pyramid.

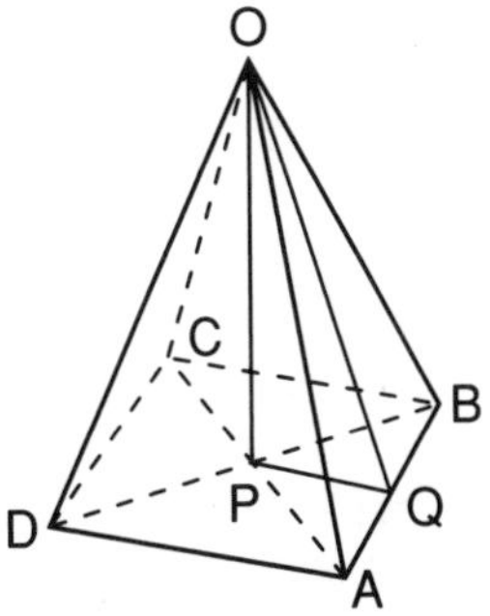

In Fig. 33, let P be the intersection of the diagonals of the base.

Fig. 33.

Join P to the vertex O.

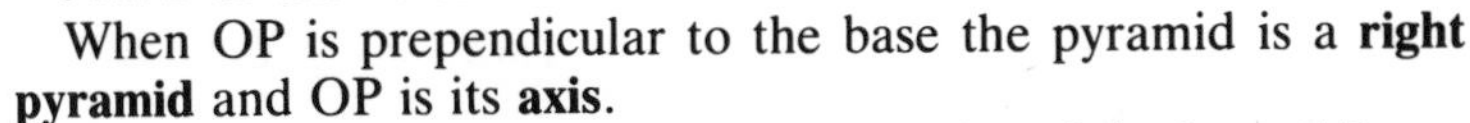

When OP is prependicular to the base the pyramid is a **right pyramid** and OP is its **axis**.

Let Q be the mid-point of one of the sides of the base AB.

Join PQ and OQ.

Then PQ and OQ are perpendicular to AB (Theorem 11).

It will be noticed that OPQ represents a plane, imagined within the pyramid but not necessarily the surface of a solid.

Then by the definition in section 20, the angle OQP represents the angle between the plane of the base and the plane of the side OAB.

Clearly the angles between the other sides and the base will be equal to this angle.

*Note* This angle must not be confused with angle OBP which students sometimes take to be the angle between a side and the base.

***Sections of right pyramids***
If sections are made parallel to the base, and therefore at right angles to the axis, they are of the same shape as the base, but of course smaller and similar.

## (3) Solids with curved surfaces

The surfaces of all the solids considered above are plane surfaces. There are many solids whose surfaces are either entirely curved or partly plane and partly curved. Three well-known ones can be mentioned here, the cylinder, the cone and the sphere. Sketches of two of these are shown below in fig. 34(a) and (b).

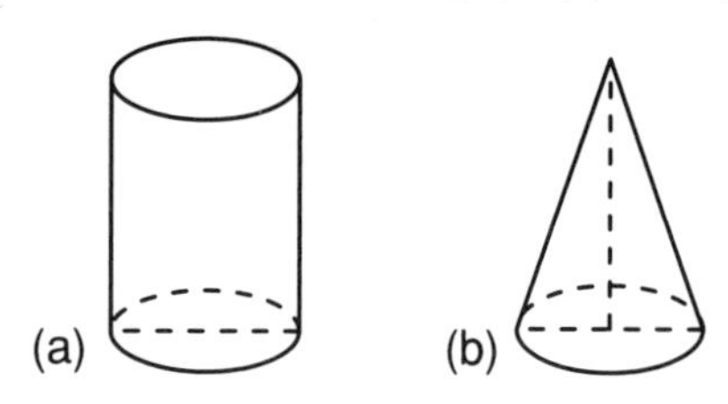


**Fig. 34.**

(*a*) The **cylinder** (Fig. 34(a)). This has two bases which are equal circles and a curved surface at right angles to these. A cylinder can be easily made by taking a rectangular piece of paper and rolling it round until two ends meet. This is sometimes called a circular prism.

(*b*) The **cone** (Fig. 34(b)). This is in reality a pyramid with a circular base.

(*c*) The **sphere**. A sphere is a solid such that any point on its surface is the same distance from a point within, called the centre. Any section of a sphere is a circle.

## 24 Angles of elevation and depression

The following terms are used in practical applications of geometry and trigonometry.

### (a) Angle of elevation

Suppose that a surveyor, standing at O (Fig. 35) wishes to determine the height of a distant tower and spire. His first step would be to place a telescope (in a theodolite) horizontally at O. He would then rotate it in a **vertical plane** until it pointed to the top of the spire. The angle through which he rotates it, the angle POQ, in Fig. 35 is called the **angle of elevation** or the **altitude** of P.

Sometimes this is said to be the angle **subtended** by the building at O.

***Altitude of the sun***

The altitude of the sun is in reality the angle of elevation of the sun. It is the angle made by the sun's rays, considered parallel, with the horizontal at any given spot at a given time.

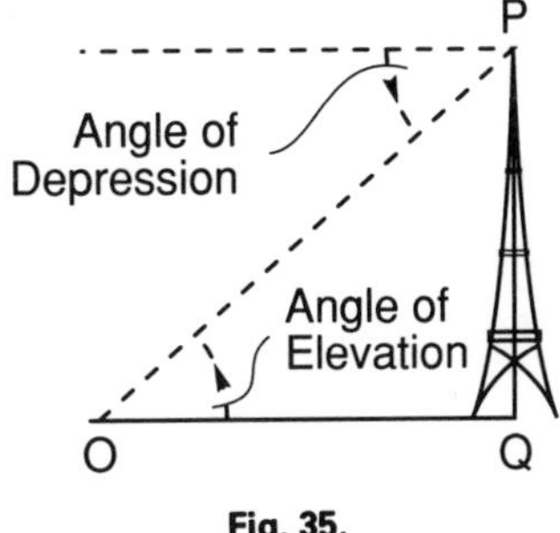


**Fig. 35.**

### (b) Angle of depression

If at the top of the tower shown in Fig. 35, a telescope were to be rotated from the horizontal until it pointed to an object at O, the angle so formed is called the **angle of depression**.

# 2

# Using your Calculator

## 25 Introduction

It is assumed in this book that you have access to a calculator. If this is not the case then you will need to be able to use logarithms or a slide rule instead.

Ideally you need a scientific calculator, that is one which has keys labelled *sin*, *cos* and *tan*, but it is possible to complete the work in this book even if you only have the simplest calculator.

The first thing that you must be aware of is that not all calculators work in the same way. In fact two different calculators can give different results for the same calculation! This can be disconcerting unless you realise what is happening. The two main differences are explained in the next section.

## 26 Arithmetic or algebraic calculators

Calculators are either designed to use **arithmetic** logic or **algebraic** logic.

*Example*: $2 + 3 \times 4$ could equal 20 or 14 depending what rules you use to decide the order of doing the addition and the multiplication.

If you carry out the sum as it is written, you would do the $2 + 3$ first to give 5 and then multiply this by 4 to give 20 as the result. This is known as using everyday or **arithmetic** logic.

On the other hand you have probably been brought up to use the rule: *do multiplications and divisions before additions and subtractions* in which case you would do $3 \times 4$ first giving 12 and then add the 2 to give 14 as the result. This is known as using **algebraic** logic.

Most scientific calculators are designed to use algebraic logic. Where brackets are not used a definite order of priority is given to the various operations. Firstly powers are carried out, then divisions followed by multiplications, subtractions and finally additions.

*Example*: Using algebraic logic $4 \times 5 + 6 \div 2 = 20 + 3 = 23$

Many of the simpler four rule calculators are designed to use everyday logic. The operations are carried out in the order that you press the keys.

*Example*: Using everyday logic
$4 \times 5 + 6 \div 2 = 20 + 6 \div 2 = 26 \div 2 = 13$

It is therefore vitally important to establish whether your calculator uses everyday or algebraic logic.

## 27 Rounding or truncating calculators

The second major difference between calculators is whether they are designed to round up any unseen figures or whether these are simply ignored.

Most calculators can display up to eight figures, although some of the more expensive ones may have ten or even twelve. However the actual calculations may be carried out using more than this number of figures.

*Example*: The result for $2 \div 3$ could be shown as 0.6666666 or 0.6666667.

In the first case, although $2 \div 3$ is 0.6666666666666 . . . , the calculator has simply cut off or truncated the result using the first eight figures only. In the second case the calculator has looked at the ninth figure and because this was five or more has rounded the previous 6 up to 7.

The majority of scientific calculators do round up whereas many of the cheaper four rule calculators simply ignore any figures which cannot be displayed.

*Exercise*: What does your calculator show for $1 \div 3 \times 3$?

On a calculator which rounds the result will be shown as 1 which is what we would have expected. However on a calculator which truncates the result will be shown as 0.9999999, since $1 \div 3 =$ 0.3333333 and when this is multiplied by 3 it becomes 0.9999999.

## 28 Differing calculator displays

Most calculators display the numbers from the right-hand end of the display. However when the result is a negative number the − sign may be separated from the number and appear at the left hand end of the display, or be immediately after the number at the right hand end of the display. On some calculators it will appear immediately in front of the number which is where we would have expected it to be.

On many scientific calculators very small numbers and very large numbers will be displayed using standard form or scientific notation. In these cases a number such as 0.000237, which in standard form can be written as $2.37 \times 10^{-4}$, will actually appear on the display as [2.37 −4].

## 29 Using your calculator for simple calculations

For straightforward arithmetic calculations you simply type in the calculation as it is written but you will need to remember to press the equals key to see the result.

*Examples*: Find $3 + 4$, $3 - 4$, $3 \times 4$, and $3 \div 4$
Type $3 + 4 =$ and you should find 7 is displayed.
Type $3 - 4 =$ and you should get −1 or 1−
Type $3 \times 4 =$ and you should get 12
Type $3 \div 4 =$ and you should get 0.75.

*Note* that providing that you finished your sequence of key presses with the = sign there was no need to clear away the last result before starting the next calculation.

## 30 The clear keys

Many calculators have designated keys for ON and OFF. The ON key often incorporates a clear function which will be shown as ON/C. When in addition there is a CE key to clear the last entry, the ON/C key usually clears the whole machine including the memories. If there is no CE key the ON/C key only acts as a clear last entry key. On some calculators these two keys are called AC for clear all and just C for clear last entry.

The CE key is used if you make a mistake in keying in a number. For example if you type 34 + 21 when you meant 34 + 12 you will need to press the CE key to get rid of the 21 before retyping the 12.

*Example*: Find 34 + 12
Switch on by pressing ON/C or AC. The 0 should appear.
Type 34 + 21, now press the CE key to remove the 21
Type 12 = and you should find that the result is 46.

*Note* if you press the wrong operation key you may be able to correct it by simply retyping the correct key immediately, but beware with some calculators both operations may well be used.

*Example*: Find 41 × 12
Type 41 + × 12 = and you will usually get 492.

*Note* if you had tried to press the CE key to clear the incorrect + sign you are likely to have got rid of the 41 as well.

It is always a good idea to precede any calculation by pressing the ON/C or all clear key and to remember to finish a calculation or part calculation by pressing the = key.

## 31 Handling minus signs and negative numbers

On a calculator it is necessary to distinguish between the operation of subtraction and the use of a − sign to indicate a negative number. The latter is usually done by having a change sign or +/− key.

When you want to enter a negative number you have to key in the number first and then press the +/− key to make it negative.

*Example*: Find −3 × −4
Type 3 [+/−] × 4 [+/−] = and you should get 12.

*Note* if you had used the subtraction key and tried −3 × −4 = you would probably have got −7. This is because the second − overrides the × sign, and the −3 was taken as 0 − 3, so the calculation was actually −3 − 4 = −7.

If your calculator does not have a +/− key then you may have some difficulty in doing calculations involving negative numbers. Most are possible by using the memory, see later in this chapter, or by doing some of the calculations on paper.

## 32 Calculations involving brackets

Many scientific calculators allow the direct use of brackets. If you have this facility most calculations can be entered exactly as they are written.

*Example*: Find (34 + 42) × (25 − 17)
Type (34 + 42) × (25 − 17) = and you should get 7.

If you do not have brackets on your calculator then you will either need to work out each bracket separately, noting down the result each time, and then combine the two separate results, or you will have to use the memory facility.

*Example*: Find (43 − 28) × (51 + 67) ÷ (19 + 43)
Find 43 − 28 = i.e. 15 and note the result.
Find 51 + 73 = i.e. 124 and note the result.
Find 19 + 43 = i.e. 62 and note the result.
Now type 15 × 124 ÷ 62 = and you should get 30 as the result.

When the calculations do not involve brackets it is important that you know whether your calculator uses algebraic logic or not. In 24 + 17 × 53 the normal convention is to do the multiplication first. With an algebraic calculator this is done automatically. With a non algebraic calculator it will be necessary to find 17 × 53 first and then add the 24.

## 33 Using the memory

Most calculators including many of the cheapest have some form of store or memory facility. These vary from one calculator to another. In some it is simply a store (STO or Min) which can be used to hold or recall (RCL or MR) one number, or be updated with a different number. In others it is possible to add (M +) or subtract (M −) additional numbers and then to recall the sum. Some calculators have more than one store which allow several numbers to be stored simultaneously.

To put a number into the store or memory you simply type in the number followed by pressing the STO or M + key. With STO the new number replaces the old number, whereas with M + the new number is added to the number which is already in the memory.

With M + it is therefore important to know whether the memory was originally empty or not. This can be done by using the ON/C key, if it clears the memory, or by using the CM key or R.CM key twice. The R on the R.CM key recalls the number in the memory and pressing it a second time then clears the memory.

Where a calculator has both an M + and an Min (or STO) key, but no clear memory facility, typing 0 followed by Min has the same effect as clearing the memory.

*Example 1*: Find 72 ÷ (23 + 13)
First make sure the memory is empty.
Check by pressing MR (or RCL) which should give 0.
Type 23 + 13 = followed by M + (or STO)
Type 72 ÷ MR (or RCL) = and the result should be 2.

The advantage of using the memory with long numbers, rather than writing down intermediate results, means that you are less likely to make errors transcribing the figures.

*Example 2a*: Find 6 × £54 + 3 × £27 + 8 × £19
Clear the memory.
Type 6 × 54 = and then M +
3 × 27 = and then M +
8 × 19 = and then M +
Then press MR to give the result £557.

*Note* it is important to remember to press the = key after each separate calculation otherwise only the second figure in each case will be added to the memory.

Where the calculator only has a store (STO) key the key presses in Example 2a are a little more complicated. The corresponding set of key presses is shown in Example 2b below.

*Example 2b*: Find 6 × £54 + 3 × £27 + 8 × £19
Clear the memory.
Type 6 × 54 = and then STO
Type 3 × 27 + RCL = and the STO
Type 8 × 19 + RCL = and this gives £557 as before.

Using the memory facility on a calculator efficiently takes quite a lot of practice. The instructions which come with your calculator usually give a number of examples and illustrate a variety of possibilities.

## 34 Using other mathematical functions

Even the simplest calculator often has at least a square root ($\sqrt{\ }$) key. Scientific calculators will also have a reciprocal (1/X) key, a $y^x$ or $x^y$ key for finding powers and roots of numbers, keys for finding the trigonometric functions of SIN, COS and TAN and their inverses together with a natural logarithm (LN) key and a base 10 logarithm (LOG) key.

The examples which follow show how some of the keys are used. If you look carefully at the keyboard of your calculator you will find that some of the above mathematical symbols are on the key itself, some are in a second colour on the key, and some are directly above the key. When the symbol is the *only* one on the key you simply press that key.

*Example 1*: Find the square root of 625, where $\sqrt{\ }$ is the only function.

Type 625$\sqrt{\ }$ which will give the result 25 directly.

Where there are two symbols on, or two symbols above, or one on and one above the key, you will find that your calculator has a special key (usually in the top left hand corner) called the INV or 2ND FN key. Pressing this key before the required key will activate the second function.

*Example 2*: Find the reciprocal of 25, where 1/X is the second function.

Type 25 INV 1/X which should give the result 0.04

*Example 3*: Find $2^5$ where there is a $y^x$ (or a $x^y$) key.

Type 2 $y^x$ 5 = which should give the result 32.

For a full explanation of all the keys on your particular calculator you will need to consult the maker's handbook supplied with your machine.

## 35 Functions and their inverses

If your calculator has an INV (or 2nd Function) key then many of the keys will have two distinct uses. The first is obtained by simply pressing the particular key, whilst the second is obtained by pressing the INV (or 2nd function) key first and then pressing the key. Often, but not always, the two functions are related. For example, if pressing the key gives the square root of the number on display then pressing INV first and then the square root key is likely to give the square of the number on the display. **Squaring** is the opposite or the inverse of finding a square root.

*Example*: Find the $\sqrt{16}$ and then show that $4^2$ is 16.
Type 16 and press the $\sqrt{\ }$ key. The result should be 4.
Now press the INV key followed by the $\sqrt{\ }$ key again.
You should have 16 on the display again.
Now press the INV key followed by the $\sqrt{\ }$ key again.
You should now have 256 (i.e. the square of 16).

You may have come across the fact that $\log_{10}3 = 0.4771 \ldots$
We can write the inverse of this statement as: $10^{0.4771} = 3$
In other words the inverse of finding the log of a number is raising the result to the power of 10.

*Example*: Find the $\log_{10}2$ and then show that $10^{0.3010}$ is 2.
Type 2 and press the LOG key.
The result should be $0.3010 \ldots$
Now press the INV key followed by the LOG key again.
You should have 2 on the display again.

On some of the keys the two functions are unrelated. For example pressing the key might give 1/X, whereas pressing INV and then the key might give you x! (factorial x).

## 36 Changing degrees to degrees, minutes and seconds

We saw, on pages 6– 7, that in ancient times each degree was subdivided into 60 minutes and each minute was further subdivided into 60 seconds. For many calculations, and especially with the introduction of the calculator, it is often more convenient to work with angles in degrees and decimals of a degree rather than with degrees, minutes and seconds. You may be lucky and have a calculator which has a DMS → DD (° ′ ″ →) key. This will allow you to do this conversion using a single key press.

*Example 1*: Change 24° 30′ to an angle using decimals of a degree.
Type in 24.30 and pres the DMS → DD key.
The display should show 24.5°

*Note*, 30′ is 0.5 of a degree.

*Example 2*: Change 24° 15′ 37″ to an angle using decimals of a degree.
Type in 24.1537 and press the DMS → DD key.
The display should show 24.260278°.

*Note*, when entering the original number the figures before the decimal point are the whole number of degrees, the first two figures after the decimal point are the number of minutes, and the next two figures the number of seconds.

To change an angle in degrees and decimals of degree to degrees, minutes and seconds you must use the INV (or 2nd function) key before pressing the DMS → DD key.

*Example 3*: Change 24.5° to an angle using degrees minutes and seconds.
Type in 24.5, press the INV key, then the DMS → DD key.
The display should show 24.3.

*Note*, the .3 represents 30 minutes.

*Example 4*: Change 24.18° to an angle using degrees minutes and seconds.
Type in 24.18, press the INV key, then the DMS → DD key.
The display should show 24.1048.

*Note*, the .1048 represents 10 minutes and 48 seconds.

## 37 Changing degrees to radians

We also saw on pages 7 and 8, that we could use radians or grades as alternative ways of measuring angles.

You may be lucky and have a calculator which has a DRG → key. This will allow you to convert degrees to radians or to grades.

*Example 1*: Change 29° to an angle in radians.
Type in 29, press the INV key, and then the DRG → key.
The display should show 0.5061455 radians.
Now press the INV key, and then the DRG → key again.
The display should show 32.222222 grades.
Now press the INV key, and then the DRG → key again.

The display should show 29° which is the angle in degrees.

*Note*, 1 radian is about 57.3°, so 29 degrees is just over half a radian.

*Example 2*: Change 2 radians to an angle in degrees.
Press the DRG → key.
The display should show that you are now in radian mode.
Type in 2, press the INV key, and then the DRG → key.
The display should show 127.32395 grades.
Now press the INV key, and then the DRG → key again.
The display should show 114.59156°, the angle in degrees.

*Note*, 1 radian is about 57.3°, so 2 radians is just over 114 degrees.

## 38 Finding trigonometric functions

If you have a scientific calculator you will have keys labelled SIN, COS and TAN. These are the trigonometric functions which you will meet in the next chapter. This and the next section are merely intended to show you how to find them and their inverses. In each case an angle is entered and a number corresponding to the angle is found.

For the examples which follow make sure your calculator is in degree mode (i.e. it is not in radian or grad mode.)

*Example 1*: Find sin 30°
Type 30 and press the SIN key.
The display should show 0.5.

*Note*, the **sin** of an angle should lie between −1 and +1.

*Example 2*: Find cos 130°
Type 130 and press the COS key.
The display should show −0.6427876.

*Note*, the **cos** of an angle should lie between −1 and +1.

*Example 3*: Find tan 230°
Type 230 and press the TAN key.
The display should show 1.1917536.

*Note*, the **tan** of an angle can take any value.

## 39 Finding inverse trigonometric functions

Here we are trying to find the angle which corresponds to a particular trigonometric ratio.

*Example 1*: Find inverse sin 0.5, also written as $\sin^{-1}0.5$
Type 0.5, press the INV key, then the SIN key.
The display should show 30°.

*Example 2*: Find inverse cos 0.5, also written as $\cos^{-1}0.5$s
Type 0.5, press the INV key, then the COS key.
The display should show 60°.

*Example 3*: Find inverse tan 1.35 also written as $\tan^{-1}1.35$
Type 1.35, press the INV key, then the TAN key.
The display should show 53.471145°.

# 3

# The Trigonometrical Ratios

## 40 The tangent

One of the earliest examples that we know in history of the practical applications of geometry was the problem of finding the height of one of the Egyptian pyramids. This was solved by Thales, the Greek philosopher and mathematician who lived about 640 BC to 550 BC. For this purpose he used the property of similar triangles which is stated in section 15 and he did it in this way.

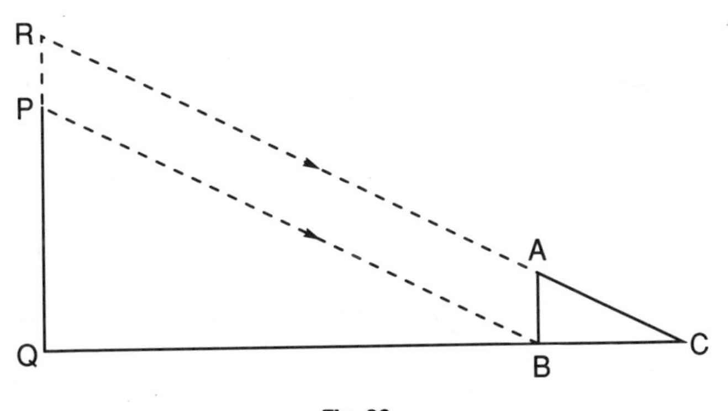


**Fig. 36.**

He observed the length of the shadow of the pyramid and, at the same time, that of a stick, AB, placed vertically into the ground at the end of the shadow of the pyramid (Fig. 36). QB represents the length of the shadow of the pyramid, and BC that of the stick. Then he said 'The height of the pyramid is to the length of the stick, as the length of the shadow of the pyramid is to the length of the shadow of the stick.'

i.e. in Fig. 36, $$\frac{PQ}{AB} = \frac{QB}{BC}.$$

Then QB, AB, and BC being known we can find PQ.

We are told that the king, Amasis, was amazed at this application of an abstract geometrical principle to the solution of such a problem.

The principle involved is practically the same as that employed in modern methods of solving the same problem. It is therefore worth examining more closely.

We note first that it is assumed that the sun's rays are parallel over the limited area involved; this assumption is justified by the great distance of the sun.

In Fig. 36 it follows that the straight lines RC and PB which represent the rays falling on the tops of the objects are parallel.

Consequently, from Theorem 2(*a*), section 9,

$$\angle PBQ = \angle ACB$$

These angles each represent the altitude of the sun (section 24).

As $\angle$s PQB and ABC are right angles

$\triangle$s PQB, ABC are similar.

$$\therefore \quad \frac{PQ}{QB} = \frac{AB}{BC}$$

or as written above $$\frac{PQ}{AB} = \frac{QB}{BC}.$$

The solution is independent of the length of the stick AB because if this be changed the length of its shadow will be changed proportionally.

We therefore can make this important general deduction.

*For the given angle ACB the ratio* $\frac{AB}{BC}$ *remains constant whatever the length of AB.*

This ratio can therefore be calculated beforehand whatever the size of the angle ACB. If this be done there is no necessity to use the stick, because knowing the angle and the value of the ratio, when we have measurred the length of QB we can easily calculate

PQ. Thus if the altitude were found to be 64° and the value of the ratio for this angle had been previously calculated to be 2.05, then we have

$$\frac{PQ}{QB} = 2.05$$

and $$PQ = QB \times 2.05.$$

## 41 Tangent of an angle

The idea of a constant ratio for every angle is vital, so we will examine it in greater detail.

Let POQ (Fig. 37) be any acute angle. From points A, B, C on one arm draw perpendiculars AD, BE, CF to the other arm. These being parallel,

$\angle$s OAD, OBE, OCF are equal (Theorem 2(*a*))

and $\angle$s ODA, OEB, OFC are right-$\angle$s.

$\therefore$ $\triangle$s AOD, BOE, COF are similar.

$$\therefore \quad \frac{AD}{OD} = \frac{BE}{OE} = \frac{CF}{OF} \quad \text{(Theorem 10, section 15)}$$

Similar results follow, no matter how many points are taken on OQ.

$\therefore$ *for the angle* POQ *the ratio of the perpendicular drawn from a point on one arm of the angle to the distance intercepted on the other arm is constant.*

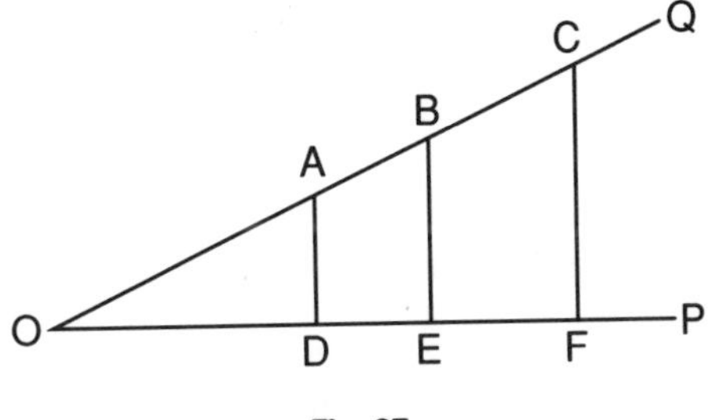


**Fig. 37.**

This is true for any angle; each angle has its own particular ratio and can be identified by it.

This constant ratio is called the **tangent** of the angle.

The name is abbreviated in use to **tan**.

Thus for ∠POQ above we can write

$$\tan \text{POQ} = \frac{\text{AD}}{\text{OD}}.$$

## 42 Right-angled triangles

Before proceeding further we will consider formally by means of the tangent, the relations which exist between the sides and angles of a right-angled triangle.

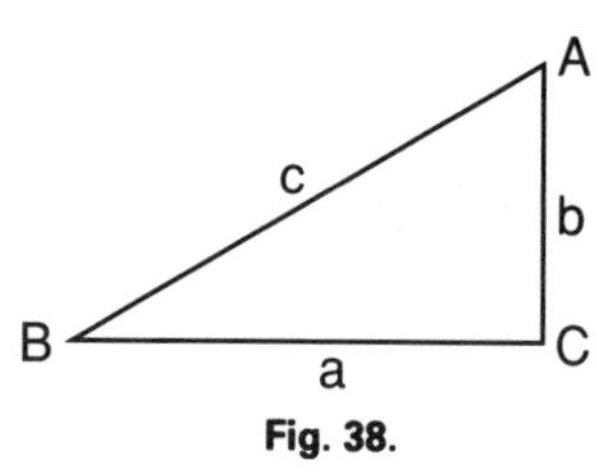


**Fig. 38.**

Let ABC (Fig. 38) be a right-angled triangle.

Let the sides opposite the angles be denoted by

a (opp. A), b (opp. B), c (opp. C).

(This is a general method of denoting sides of a right-angled △.)

Then, as shown in section 41:

$$\tan B = \frac{AC}{BC} = \frac{b}{a}$$

$$\therefore \quad a \tan B = b$$

and

$$a = \frac{b}{\tan B}.$$

Thus any one of the three quantities a, b, tan B can be determined when the other two are known.

## 43 Notation for angles

(*a*) As indicated above we sometimes, for brevity, refer to an angle by using only the middle letter of the three which define the angle.

Thus we use tan B for tan ABC.

This must not be used when there is any ambiguity as, for example, when there is more than one angle with its vertex at the same point.

(*b*) When we refer to angles in general we frequently use a Greek letter, usually $\theta$ (pronounced 'theta') or $\phi$ (pronounced 'phi') or $\alpha$, $\beta$ or $\gamma$ (alpha, beta, gamma).

## 44 Changes in the tangent in the first quadrant

In Fig. 39 let OA a straight line of **unit** length rotate from a fixed position on OX until it reaches OY, a straight line perpendicular to OX.

From O draw radiating lines to mark 10°, 20°, 30°, etc.

From A draw a straight line AM perpendicular to OX and let the radiating lines be produced to meet this.

Let OB be any one of these lines.

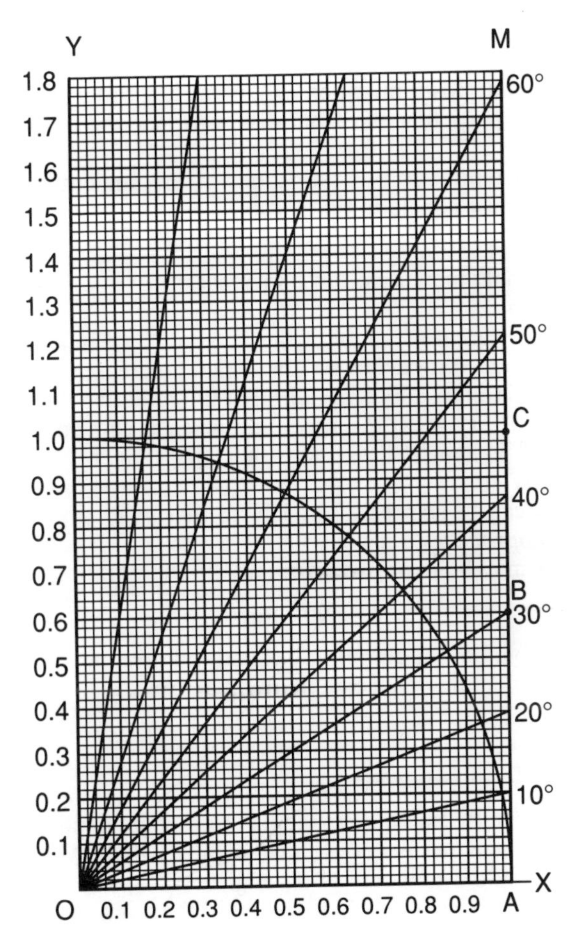


**Fig. 39.**

Then $$\tan \text{BOA} = \frac{\text{BA}}{\text{OA}}.$$

Since OA is of unit length, then the length of BA, on the scale selected, will give the actual value of tan BOA.

Similarly the tangents of other angles 10°, 20°, etc. can be read off by measuring the corresponding intercept on AM.

If the line OC corresponding to 45° be drawn then ∠ACO is also 45° and AC equals OA (Theorem 3, section 11).

$$\therefore \quad \text{AC} = 1$$
$$\therefore \quad \tan 45° = 1$$

At the initial position, when OA is on OX the angle is 0°, the length of the perpendicular from A is zero, and the tangent is also zero.

From an examination of the values of the tangents as marked on AM, we may conclude that:

(1) tan 0° is 0;
(2) as the angle increases, tan $\theta$ increases;
(3) tan 45° = 1;
(4) for angles greater than 45°, the tangent is greater than 1;
(5) as the angle approaches 90° the tangent increases very rapidly. When it is almost 90° it is clear that the radiating line will meet AM at a very great distance, and when it coincides with OY and 90° is reached, we say that the tangent has become infinitely great.

This can be expressed by saying that *as $\theta$ approaches 90°, tan $\theta$ approaches infinity*.

This may be expressed formally by the notation

when $$\theta \to 90°, \tan \theta \to \infty.$$

The symbol $\infty$, commonly called infinity, means a number greater than any conceivable number.

## 45 A table of tangents

Before use can be made of tangents in practical applications and calculations, it is necessary to have a table which will give with great accuracy the tangents of all angles which may be required. It must also be possible from it to obtain the angle corresponding to a known tangent.

A rough table could be constructed by such a practical method as is indicated in the previous paragraph. But results obtained in this way would not be very accurate.

By the methods of more advanced mathematics, however, these values can be calculated to any required degree of accuracy. For elementary work it is customary to use tangents calculated correctly to four places of decimals. Such a table can be found at the end of this book.

A small portion of this table, giving the tangents of angles from 25° to 29° inclusive is given below, and this will serve for an explanation as to how to use it.

*Natural Tangents*

| Degrees | 0′ | 6′ 0.1° | 12′ 0.2° | 18′ 0.3° | 24′ 0.4° | 30′ 0.5° | 36′ 0.6° | 42′ 0.7° | 48′ 0.8° | 54′ 0.9° | Mean Differences 1 | 2 | 3 | 4 | 5 |
|---|---|---|---|---|---|---|---|---|---|---|---|---|---|---|---|
| 25 | 0.4663 | 4684 | 4706 | 4727 | 4748 | 4770 | 4791 | 4813 | 4834 | 4856 | 4 | 7 | 11 | 14 | 18 |
| 26 | 0.4877 | 4899 | 4921 | 4942 | 4964 | 4986 | 5008 | 5029 | 5051 | 5073 | 4 | 7 | 11 | 15 | 18 |
| 27 | 0.5095 | 5117 | 5139 | 5161 | 5184 | 5206 | 5228 | 5250 | 5272 | 5295 | 4 | 7 | 11 | 15 | 18 |
| 28 | 0.5317 | 5340 | 5362 | 5384 | 5407 | 5430 | 5452 | 5475 | 5498 | 5520 | 4 | 8 | 11 | 15 | 18 |
| 29 | 0.5543 | 5566 | 5589 | 5612 | 5635 | 5658 | 5681 | 5704 | 5727 | 5750 | 4 | 8 | 12 | 15 | 19 |

(1) The first column indicates the angle in degrees.

(2) The second column states the corresponding tangent.
Thus $\tan 27° = 0.5095$

(3) If the angle includes minutes we must use the remaining columns.

(*a*) If the number of minutes is a multiple of 6 the figures in the corresponding column give the decimal part of the tangent. Thus tan 25° 24′ will be found under the column marked 24′. From this we see

$$\tan 25° 24' = 0.4748.$$

On your calculator check that tan25.4° is 0.4748 correct to 4 decimal places, i.e. enter 25.4, and press the TAN key.

If you are given the value of the tan and you want to obtain the angle, you should enter 0.4748, press the INV key and then press the TAN key, giving 25.40° as the result correct to 2 decimal places.

(*b*) If the number of minutes is not an exact multiple of 6, we use the columns headed 'mean differences' for angles which are 1, 2, 3, 4, or 5 minutes more than the multiple of 6.

Thus if we want tan 26° 38′, this being 2′ more than 26° 36′, we look under the column headed 2 in the line of 26°. The difference is 7. This is added to tan 26° 36′, i.e. 0.5008.

Thus $$\tan 26° 38' = 0.5008 + .0007 = 0.5015.$$

An examination of the first column in the table of tangents will show you that as the angles increase and approach 90° the tangents increase very rapidly. Consequently for angles greater than 45° the whole number part is given as well as the decimal part. For angles greater than 74° the mean differences become so large and increase so rapidly that they cannot be given with any degree of accuracy.

## 46 Examples of the uses of tangents

We will now consider a few examples illustrating practical applications of tangents. The first is suggested by the problem mentioned in section 24.

*Example 1*: At a point 168 m horizontally distant from the foot of a church tower, the angle of elevation of the top of the tower is 38° 15′.

Find the height above the ground of the top of the tower.

In Fig. 40 PQ represents the height of P above the ground. We will assume that the distance from O is represented by OQ. Then ∠POQ is the angle of elevation and equals 38.25°.

$$\therefore \quad \frac{PQ}{OQ} = \tan 38.25°$$

$$\begin{aligned} \therefore \quad PQ &= OQ \times \tan 38.25° \\ &= 168 \times \tan 38.25° \\ &= 168 \times 0.7883364 \\ &= 132.44052 \\ \therefore \quad PQ &= 132 \text{ m approx.} \end{aligned}$$

On your calculator the sequence of key presses should be:
38.25 TAN × 168 =, giving 132.44052, or 132.44 m as the result.

*Example 2*: A man, who is 168 cm in height, noticed that the length of his shadow in the sun was 154 cm. What was the altitude of the sun?

In Fig. 41 let PQ represent the man and QR represent the shadow.

Then PR represents the sun's ray and $\angle$PRQ represents the sun's altitude.

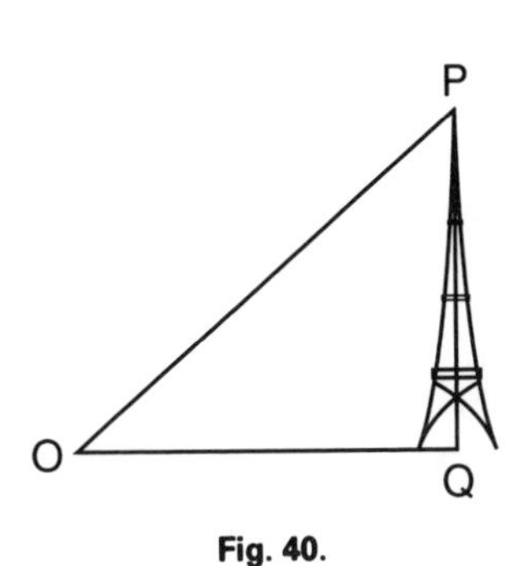


**Fig. 40.**

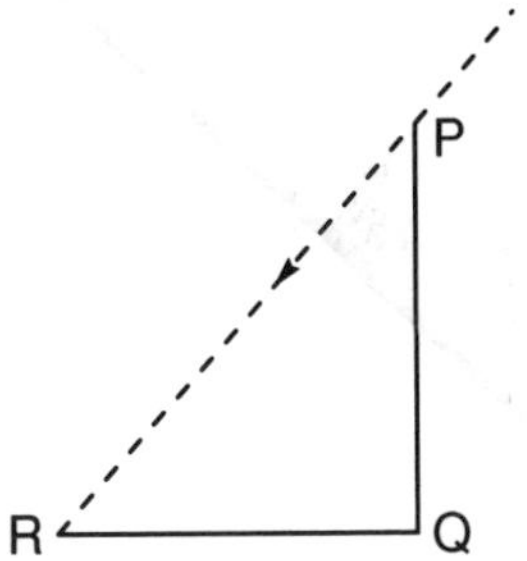

**Fig. 41.**

Now
$$\tan PRQ = \frac{PQ}{QR} = \frac{168 \text{ cm}}{154 \text{ cm}}$$
$$= 1.0909 \text{ (approx.)}$$
$$= \tan 47.49°$$
$\therefore$ the sun's altitude is 47.49° or 47° 29′

*Example 3*: Fig. 42 represents a section of a symmetrical roof in which AB is the span, and OP the rise. (P is the mid-point of AB.) If the span is 22 m and the rise 7 m find the slope of the roof (i.e. the angle OBA).

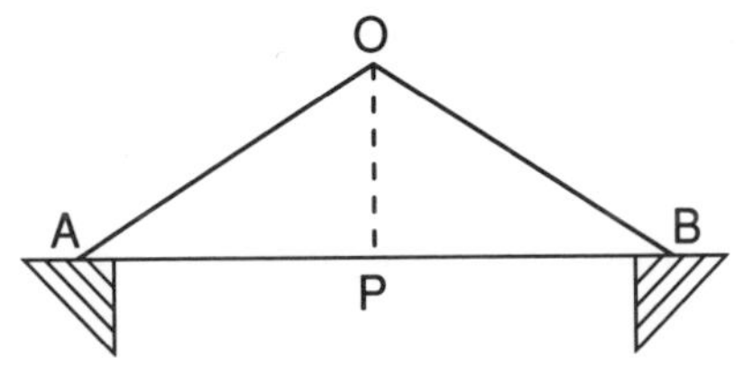


**Fig. 42.**

OAB is an isosceles triangle, since the roof is symmetrical.
$\therefore$ OP is perpendicular to AB (Theorem 3, section 11)

$$\therefore \tan OBP = \frac{OP}{PB}$$

$$= \frac{7}{11} = 0.6364 \text{ (approx.)}$$
$$= \tan 32.47° \text{ (approx.)}$$
$$\therefore \angle OBP = 32.47° \text{ or } 32° 28'$$

On your calculator the sequence of key presses should be:
7 ÷ 11 = INV TAN, giving 32.471192, or 32.47° as the result.

*Exercise 1*

**1** In Fig. 43 ABC is a right-angled triangle with C the right angle.

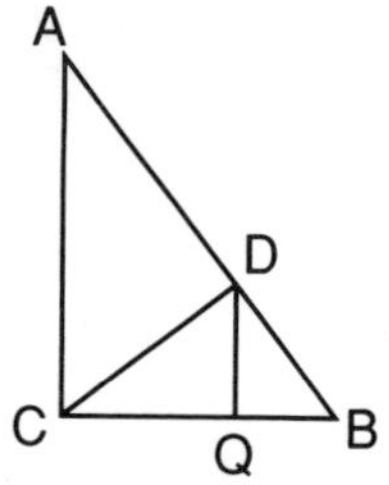

Draw CD perpendicular to AB and DQ perpendicular to CB.

Write down the tangents of ABC and CAB in as many ways as possible, using lines of the figure.

A
D
C
Q
B

**Fig. 43.**

**2** In Fig. 43, if AB is 15 cm and AC 12 cm in length, find the values of tan ABC and tan CAB.

**3** From the tables write down the tangents of the following angles:

(1) 18° (2) 43° (3) 56°
(4) 73° (5) 14° 18′ (6) 34° 48′

Check your results on your calculator, i.e. enter the angle and press the TAN key.

**4** Write down the tangents of:

(1) 9° 17′ (2) 31.75° (3) 39° 5′
(4) 52.45° (5) 64° 40′

**5** From the tables find the angles whose tangents are:

(1) 0.5452 (2) 1.8265 (3) 2.8239
(4) 1.3001 (5) 0.6707 (6) 0.2542

Check your results on your calculator, i.e. enter the number, press the INV key and then the TAN key.

**6** When the altitude of the sun is 48° 24′, find the height of a flagstaff whose shadow is 7.42 m long.

**7** The base of an isosceles triangle is 10 mm and each of the equal sides is 13 mm. Find the angles of the triangle.

**8** A ladder rests against the top of the wall of a house and makes an angle of 69° with the ground. If the foot is 7.5 m from the wall, what is the height of the house?

**9** From the top window of a house which is 1.5 km away from a tower it is observed that the angle of elevation of the top of the tower is 36° and the angle of depression of the bottom is 12°. What is the height of the tower?

**10** From the top of a cliff 32 m high it is noted that the angles of depression of two boats lying in the line due east of the cliff are 21° and 17°. How far are the boats apart?

**11** Two adjacent sides of a rectangle are 15.8 cms and 11.9 cms. Find the angles which a diagonal of the rectangle makes with the sides.

**12** P and Q are two points directly opposite to one another on the banks of a river. A distance of 80 m is measured along one bank at right angles to PQ. From the end of this line the angle subtended by PQ is 61°. Find the width of the river.

## 47 Sines and cosines

In Fig. 44 from a point A on one arm of the angle ABC, a perpendicular is drawn to the other arm.

We have seen that the ratio $\frac{AC}{BC}$ = tan ABC.

Now let us consider the ratios of each of the lines AC and BC to the hypotenuse AB.

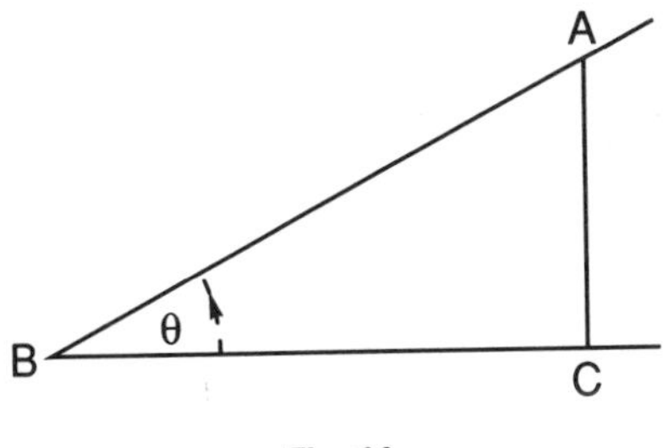


**Fig. 44.**

(1) The ratio $\frac{AC}{AB}$, i.e. the ratio of the side opposite to the angle to the hypotenuse.

This ratio is also constant, as was the tangent, for the angle ABC, i.e. wherever the point A is taken, the ratio of AC to AB remains constant.

This ratio is called the **sine of the angle** and is denoted by sin ABC.

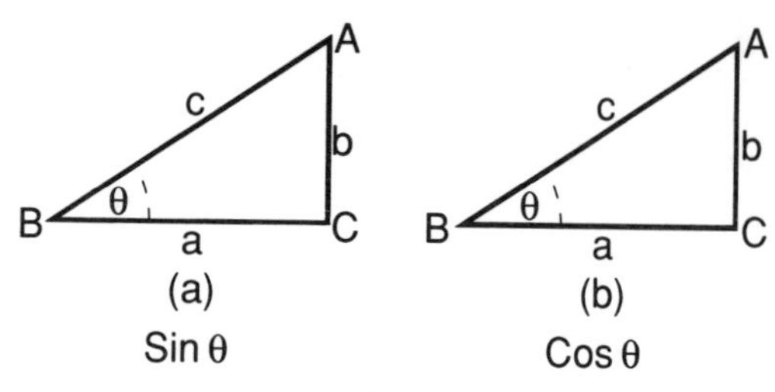


**Fig. 45.**

(2) The ratio $\frac{BC}{AB}$, i.e. the ratio of the intercept to the hypotenuse.

This ratio is also constant for the angle and is called the **cosine**. It is denoted by cos ABC.

Be careful not to confuse these two ratios. The way in which they are depicted by the use of thick lines in Fig. 45 may help you. If the sides of the $\triangle$ABC are denoted by a, b, c in the usual way and the angle ABC by θ (pronounced theta).

Then in 45(a) $\qquad \sin\theta = \frac{b}{c} \qquad (1)$

45(b) $\qquad \cos\theta = \frac{a}{c} \qquad (2)$

From (1) we get $\qquad b = c\sin\theta$
From (2) we get $\qquad a = c\cos\theta$

Since in the fractions representing sin θ and cos θ above, the denominator is the hypotenuse, which is the greatest side of the triangle, then *sin θ and cos θ cannot be greater than unity*.

## 48 Ratios of complementary angles

In Fig. 45, since $\angle C$ is a right angle.

$$\therefore \angle A + \angle B = 90°$$

$\therefore$ $\angle A$ and $\angle B$ are complementary (see section 7).

Also $\qquad \sin A = \frac{a}{c}$

and
$$\cos B = \frac{a}{c}$$

$$\therefore \quad \sin A = \cos B.$$

∴ The sine of an angle is equal to the cosine of its complement, and vice versa.

This may be expressed in the form:

$$\sin \theta = \cos (90^\circ - \theta)$$
$$\cos \theta = \sin (90^\circ - \theta).$$

## 49 Changes in the sines of angles in the first quadrant

Let a line, OA, a **unit in length**, rotate from a fixed position (Fig. 46) until it describes a quadrant, that is the ∠DOA is a right angle.

From O draw a series of radii to the circumference corresponding to the angles 10°, 20°, 30°, . . .

From the points where they meet the circumference draw lines perpendicular to OA.

Considering any one of these, say BC, corresponding to 40°.

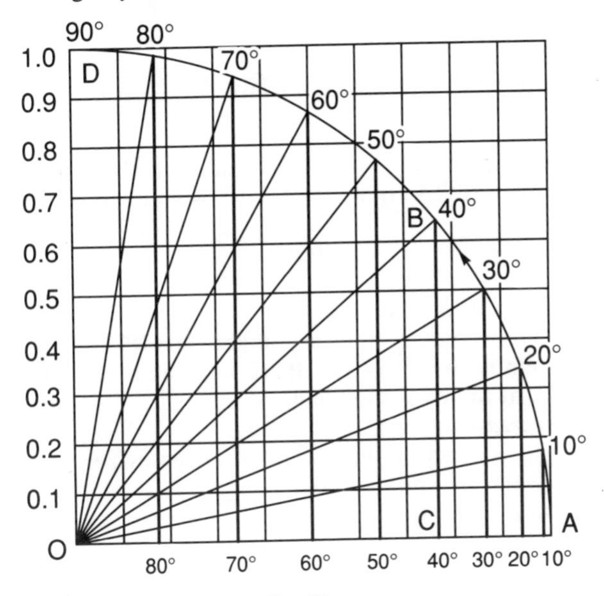


Fig. 46.

Then $$\sin \text{BOC} = \frac{\text{BC}}{\text{OB}}.$$

But OB is of unit length.

$\therefore$ BC represents the value of sin BOC, in the scale in which OA represents unity.

Consequently the various perpendiculars which have been drawn represent the sines of the corresponding angles.

Examining these perpendiculars we see that *as the angles increase from 0° to 90° the sines continually increase.*

At 90° the perpendicular coincides with the radius

$$\therefore \quad \sin 90° = 1$$

At 0° the perpendicular vanishes

$$\therefore \quad \sin 0° = 0$$

Summarising these results:

In the first quadrant

(1) $\sin 0° = 0$,
(2) as $\theta$ increases from 0° to 90°, $\sin \theta$ increases,
(3) $\sin 90° = 1$.

## 50 Changes in the cosines of angles in the first quadrant

Referring again to Fig. 46 and considering the cosines of the angles formed as OA rotates, we have as an example

$$\cos \text{BOC} = \frac{\text{OC}}{\text{OB}}$$

As before, OB is of unit length.

$\therefore$ OC represents in the scale taken, cos BOC.

Consequently the lengths of these intercepts on OA represent the cosines of the corresponding angles.

*These decrease as the angle increases.*

When 90° is reached this intercept becomes zero and at 0° it coincides with OA and is unity.

Hence in the first quadrant

(1) $\cos 0° = 1$
(2) As $\theta$ increases from 0° to 90°, $\cos \theta$ decreases,
(3) $\cos 90° = 0$.

## 51 Tables of sines and cosines

As in the case of the tangent ratio, it is necessary to compile tables giving the values of these ratios for all angles if we are to use sines and cosines for practical purposes. These have been calculated and arranged by methods similar to the tangent tables, and the general directions given in section 45 for their use will also apply to those for sines and cosines.

The table for cosines is not really essential when we have the tables of sines, for since $\cos \theta = \sin (90° - \theta)$ (see section 48) we can find cosines of angles from the sine table.

For example, if we require cos 47°, we know that

$$\begin{aligned} \cos 47° &= \sin (90° - 47°) \\ &= \sin 43°. \end{aligned}$$

$\therefore$ to find cos 47° we read the value of sin 43° in the sine table.

In practice this process takes longer and is more likely to lead to inaccuracies than finding the cosine direct from a table. Consequently separate tables for cosines are included at the end of this book.

There is one difference between the sine and cosine tables which you need to remember when you are using them.

We saw in section 50, that as angles in the first quadrant increase, *sines increase but cosines decrease*. Therefore when using the columns of mean differences for cosines these differences must be **subtracted**.

On your calculator check that sin 43° is 0.6820 correct to 4 decimal places, i.e. enter 43, and press the SIN key.

On your calculator check that cos 47° is also 0.6820 correct to 4 decimal places, i.e. enter 47, and press the COS key.

If you are given the value of the sin (or cos) and you want to obtain the angle, you should enter the number, 0.6820, press the INV key and then press the SIN (or COS) key, giving 43.00° (or 47.00°) as the result correct to 2 decimal places.

## 52 Examples of the use of sines and cosines

*Example 1*: The length of each of the legs of a pair of ladders is 2.5 m. The legs are opened out so that the distance between the feet is 2 m. What is the angle between the legs?

In Fig. 47, let AB, AC represent the legs of the ladders. These being equal, BAC is an isosceles triangle.

$\therefore$ AO the perpendicular to the base BC, from the vertex bisects the vertical angle BAC, and also the base.

$$\therefore \quad BO = OC = 1 \text{ m}$$

We need to find the angle BAC.

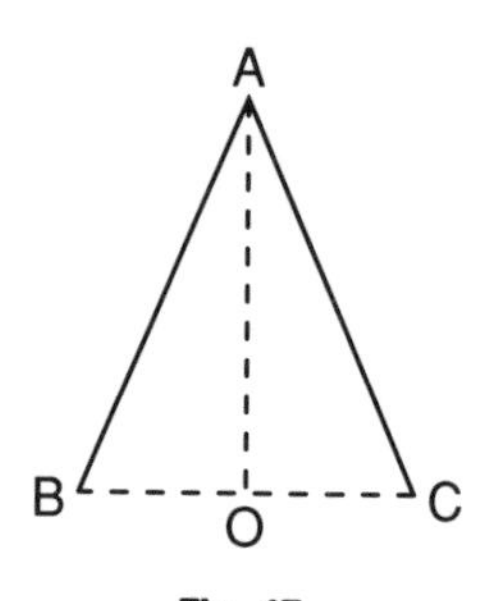


**Fig. 47.**

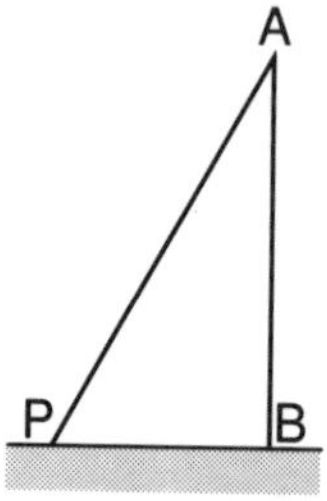

**Fig. 48.**

Now $$\sin BAO = \frac{BO}{BA}$$

$$= \frac{1}{2.5} = 0.4$$

$$= \sin 23.58° \text{ (from the tables)}$$

$$\therefore \quad \angle BAO = 23.58°$$

But $$\angle BAC = 2 \times \angle BAO$$

$$\therefore \quad \angle BAC = 2 \times 23.58°$$
$$= 47.16°$$

On your calculator the sequence of key presses should be:
  1 ÷ 2.5 = INV SIN × 2, giving 47.156357, or 47.16° as the result.

*Example 2*: A 30 m ladder on a fire engine has to reach a window 26 m from the ground which is horizontal and level. What angle, to the nearest degree, must it make with the ground and how far from the building must it be placed?

Let AB (Fig. 48) represent the height of the window at A above the ground.

Let AP represent the ladder.

To find $\angle$APB we may use its sine for

$$\text{Sin APB} = \frac{AB}{AP} = \frac{26}{30}$$

$$= 0.8667$$
$$= \sin 60.07° \text{ (from the tables)}$$
$$\therefore \quad APB = 60.07°$$
$$= 60° \text{ (to nearest degree).}$$

On your calculator the sequence of key presses should be:
26 ÷ 30 = INV SIN, giving 60.073565, or 60.07° as the result.

To find PB we use the cosine of APB

for $$\cos APB = \frac{PB}{AP}$$

$$\therefore \quad PB = AP \cos APB$$
$$= 30 \times \cos 60.07°$$
$$= 30 \times 0.4989$$
$$= 14.97$$
$$\therefore \quad PB = 15 \text{ m (approx.)}$$

On your calculator the sequence of key presses should be:
30 × 60.07 COS =, giving 14.968247, or 14.97 m as the result.

*Example 3*: The height of a cone is 18 cm and the angle at the vertex is 88°. Find the slant height.

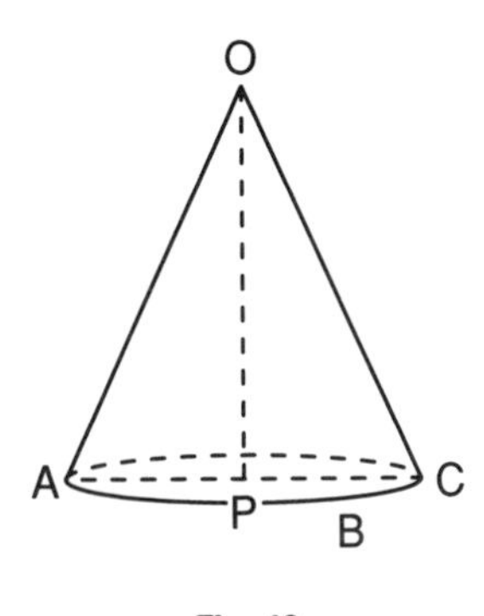

Fig. 49.

Let OABC (Fig. 49) represent the cone, the vertex being O and ABC the base.

Let the △OAC represent a section through the vertex O and perpendicular to the base.

It will be an isosceles triangle and P the centre of its base will be the foot of the perpendicular from O to the base.

OP will also bisect the vertical angle AOC (Theorem 3).

OP represents the height of the cone and is equal to 18 cm.

OC represents the slant height.

Now $$\cos POC = \frac{OP}{OC}$$

$$\therefore \quad OP = OC \cos POC$$
$$\therefore \quad OC = OP \div \cos POC$$
$$= 18 \div \cos 44°$$

$$= 18 \div 0.7193$$
$$= 25.02$$
$$\therefore \quad OC = 25 \text{ cm (approx.).}$$

On your calculator the sequence of key presses should be:
18 ÷ 44 COS =, giving 25.022945, or 25.02 cm as the result.

*Example 4*: Fig. 50 represents a section of a symmetrical roof frame. PA = 28 m, AB = 6 m, ∠OPA = 21°; find OP and OA.

(1) We can get OP if we find ∠OPB. To do this we must first find ∠APB.

$$\sin APB = \frac{AB}{AP} = \frac{6}{28} = 0.2143 = \sin 12.37°$$

$$\therefore \quad \angle OPB = \angle OPA + \angle APB$$
$$= 21° + 12.37° = 33.37°$$

Next find PB, which divided by OP gives cos OPB.

$$PB = AP \cos APB = 28 \cos 12.37°$$
$$= 28 \times 0.9768$$
$$= 27.35 \text{ (approx.)}$$

*Note* We could also use the Theorem of Pythagoras.

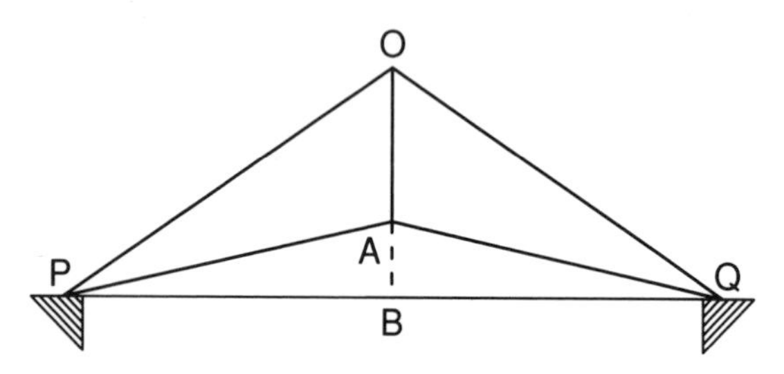


**Fig. 50.**

Now $$\frac{PB}{OP} = \cos OPB$$

$$\therefore \quad OP = PB \div \cos OPB$$
$$\therefore \quad OP = 27.35 \div \cos 33.38°$$
$$= 27.35 \div 0.8350$$
$$= 32.75$$
$$\therefore \quad OP = 32.75 \text{ m.}$$

On your calculator the sequence of key presses should be:
27.35 ÷ 33.38 COS =, giving 32.752923, or 35.75 m as the result.

(2) To find OA. This is equal to OB − AB. We must therefore find OB.

Now $$\frac{OB}{OP} = \sin OPB$$

$$\therefore \quad OB = OP \sin OPB$$
$$= 32.75 \times \sin 33.38°$$
$$= 32.75 \times 0.5502$$
$$= 18.02$$

and $$OA = OB - AB$$
$$= 18.02 - 6$$
$$= 12.02 \text{ m.}$$

On your calculator the sequence of key presses should be:
32.75 × 33.38 SIN − 6 =, giving 12.018699, or 12.02 m as the result.

*Exercise 2*

**1** Using the triangle of Fig. 43 write down in as many ways as possible (1) the sines, (2) the cosines, of ∠ABC and ∠CAB, using the lines of the figure.

**2** Draw a circle with radius 45 mm. Draw a chord of length 60 mm. Find the sine and cosine of the angle subtended by this chord at the centre.

**3** In a circle of 4 cm radius a chord is drawn subtending an angle of 80° at the centre. Find the length of the chord and its distance from the centre.

**4** The sides of a triangle are 135 mm, 180 mm, and 225 mm. Draw the triangle, and find the sines and cosines of the angles.

**5** From the tables write down the sines of the following angles:

(1) 14° 36′  (2) 47.43°  (3) 69° 17′

Check your results on your calculator,
i.e. enter the angle and press the SIN key.

**6** From the tables write down the angles whose sines are:

(1) 0.4970  (2) 0.5115  (3) 0.7906

Check your results on your calculator,
i.e. enter the number, press the INV key and then the SIN key.

**7** From the tables write down the cosines of the following angles:

(1) 20° 46′  (2) 44° 22′  (3) 62° 39′

(4) 38.83° (5) 79.27° (6) 57.38°

Check your results on your calculator,
i.e. enter the angle and press the COS key.

**8** From the tables write down the angles whose cosines are:

(1) 0.5332 (2) 0.9358 (3) 0.3546
(4) 0.2172 (5) 0.7910 (6) 0.5140

Check your results on your calculator,
i.e. enter the number, press the INV key and then the COS key.

**9** A certain uniform incline rises 10.5 km in a length of 60 km along the incline. Find the angle between the incline and the horizontal.

**10** In a right-angled triangle the sides containing the right angle are 4.6 m and 5.8 m. Find the angles and the length of the hypotenuse.

**11** In the diagram of a roof frame shown in Fig. 42, find the angle at which the roof is sloped to the horizontal when OP = 1.3 m and OB = 5.4 m.

**12** A rope 65 m long is stretched out from the top of a flagstaff 48 m high to a point on the ground which is level. What angle does it make with the ground and how far is this point from the foot of the flagstaff?

## 53 Cosecant, secant and cotangent

From the reciprocals of the sine, cosine and tangent we can obtain three other ratios connected with an angle, and problems frequently arise where it is more convenient to employ these instead of using the reciprocals of the original ratios.

These reciprocals are called the **cosecant**, **secant**, and **cotangent** respectively, abbreviated to cosec, sec and cot.

Thus

$$\operatorname{cosec} \theta = \frac{1}{\sin \theta}$$

$$\sec \theta = \frac{1}{\cos \theta}$$

$$\cot \theta = \frac{1}{\tan \theta}$$

These can be expressed in terms of the sides of a right-angled triangle with the usual construction (Fig. 51) as follows:

$$\frac{AC}{AB} = \sin\theta, \ \frac{AB}{AC} = \operatorname{cosec}\theta$$

$$\frac{BC}{AB} = \cos\theta, \ \frac{AB}{BC} = \sec\theta$$

$$\frac{AC}{BC} = \tan\theta, \ \frac{BC}{AC} = \cot\theta$$

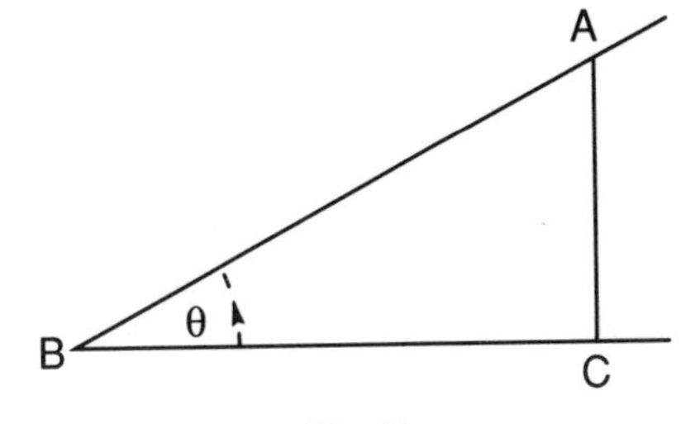


**Fig. 51.**

**Ratios of complementary angles**

In continuation of section 48 we note that:

since $$\tan ABC = \frac{AC}{BC}$$

and $$\cot BAC = \frac{AC}{BC}$$

$$\therefore \quad \tan\theta = \cot(90° - \theta)$$

or *the tangent of an angle is equal to the cotangent of its complement.*

## 54 Changes in the reciprocal ratios of angles in the first quadrant

The changes in the values of these ratios can best be examined by reference to the corresponding changes in the values of their reciprocals (see sections 44, 49 and 50 in this chapter).

The following general relations between a ratio and its reciprocal should be noted:

(*a*) *When the ratio is increasing its reciprocal is decreasing*, and vice versa.

(*b*) *When a ratio is a maximum its reciprocal will be a minimum*, and vice versa.

Consequently since the **maximum** value of the sine and cosine in the first quadrant is unity, the **minimum** value of the cosecant and secant must be unity.

(*c*) The case when a ratio is zero needs special examination.

If a number is very large, its reciprocal is very small. Conversely if it is very small its reciprocal is very large.

Thus the reciprocal of $\frac{1}{1\,000\,000}$ is 1 000 000.

When a ratio such as a cosine is decreasing until it finally becomes zero, as it does when the angle reaches 90°, the secant approaches infinity. With the notation employed in section 44 this can be expressed as follows.

As $\theta \to 90°$, $\sec \theta \to \infty$.

## 55 Changes in the cosecant

Bearing in mind the above, and remembering the changes in the sine in the first quadrant as given in section 49:

(1) cosec 0° is infinitely large,
(2) as $\theta$ *increases* from 0° to 90°, cosec $\theta$ *decreases*,
(3) cosec 90° = 1.

## 56 Changes in the secant

Comparing with the corresponding changes in the cosine we see:

(1) sec 0° = 1,
(2) as $\theta$ *increases* from 0 to 90°, sec $\theta$ *increases*,
(3) as $\theta \to 90°$, $\sec \theta \to \infty$.

## 57 Changes in the cotangent

Comparing the corresponding changes of the tan $\theta$ as given in section 44 we conclude:

(1) as $\theta \to 0°$, $\cot \theta \to \infty$,
(2) as $\theta$ *increases*, cot $\theta$ *decreases*,
(3) cot 45° = 1,
(4) as $\theta \to 90°$, $\cot \theta \to 0$.

## 58 Using your calculator for other trigonometrical ratios

Most scientific calculators do not have separate keys for cosecant (COSEC), secant (SEC) and cotangent (COT). However this is not necessary since each is simply the reciprocal of the corresponding sine, cosine and tangent values.

*Example 1*: Find the value of the cosecant of 30°
Type 30, press the SIN key and then the 1/X key.
The display should show 2.

*Note*, the cosecant of 0° does not exist since sin 0° is 0, and you cannot divide by zero.
An alternative method would be to type 1 ÷ 30 SIN =

*Example 2*: Find the value of the secant of 30°
Type 30, press the COS key and then the 1/X key.
The display should show 1.1547005.

*Note*, the secant of 90° and 270° do not exist since cos 90° and cos 270° are each 0, and you cannot divide by zero.
An alternative method would be to type 1 ÷ 30 COS =

*Example 3*: Find the value of the cotangent of 30°
Type 30, press the TAN key and then the 1/X key.
The display should show 1.7320508.

*Note*, the cotangent of 0° and 180° do not exist since tan 0° and tan 180° are each 0, and you cannot divide by zero.
An alternative method would be to type 1 ÷ 30 TAN =

## 59 Graphs of the trigonometrical ratios

In Figs. 52, 53, 54 are shown the graphs of sin $\theta$, cos $\theta$ and tan $\theta$ respectively for angles in the first quadrant. You should draw them yourself, if possible, on squared paper, obtaining the values either by the graphical methods suggested in Figs. 39 and 46 or from the tables.

## 60 Uses of other trigonometrical ratios

*Worked Examples*

*Example 1*: Find (i) cosec 37.5° and (ii) the angle whose cotan is 0.8782.

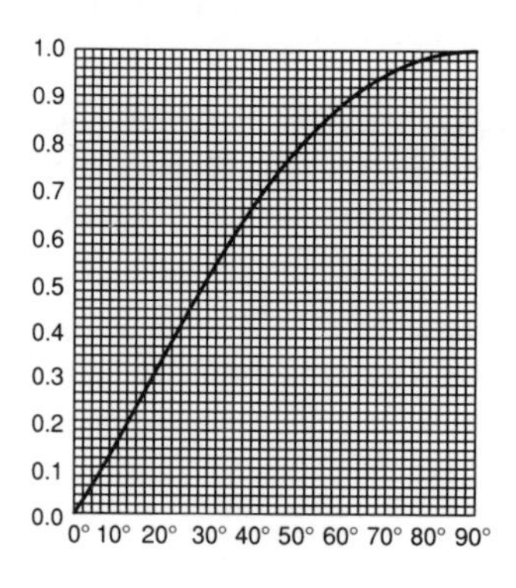


**Fig. 52.**
**Graph of *sin* θ.**

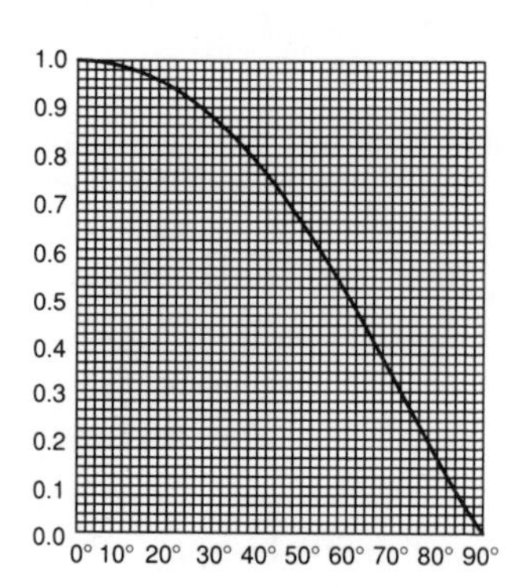


**Fig. 53.**
**Graph of *cos* θ.**

(i) cosec 37.5° = 1 ÷ sin 37.5°,
so type 1 ÷ 37.5 SIN =
The result should be 1.6426796

(ii) 1 ÷ cotan θ = tan θ,
so type 1 ÷ 0.8782 = INV TAN
The result should be 48.71°

*Example 2*: From a certain point the angle of elevation of the top of a church spire is found to be 11°. The guide book tells me that the height of the spire is 260 m. If I am on the same horizontal level as the bottom of the tower, how far am I away from it?

In Fig. 55 let AB represent the tower and spire,

$$AB = 260 \text{ m}$$

Let O be the point of observation.
We need to find OB.

Let $$OB = x$$

Then $$\frac{x}{260} = \cot 11°$$

$$\therefore \quad x = 260 \cot 11° \qquad (1)$$
$$\therefore \quad x = 260 \div \tan 11°$$
$$\therefore \quad x = 260 \div 0.1944$$
$$= 1338$$
$$\therefore \quad x = 1338 \text{ m (approx.)}.$$

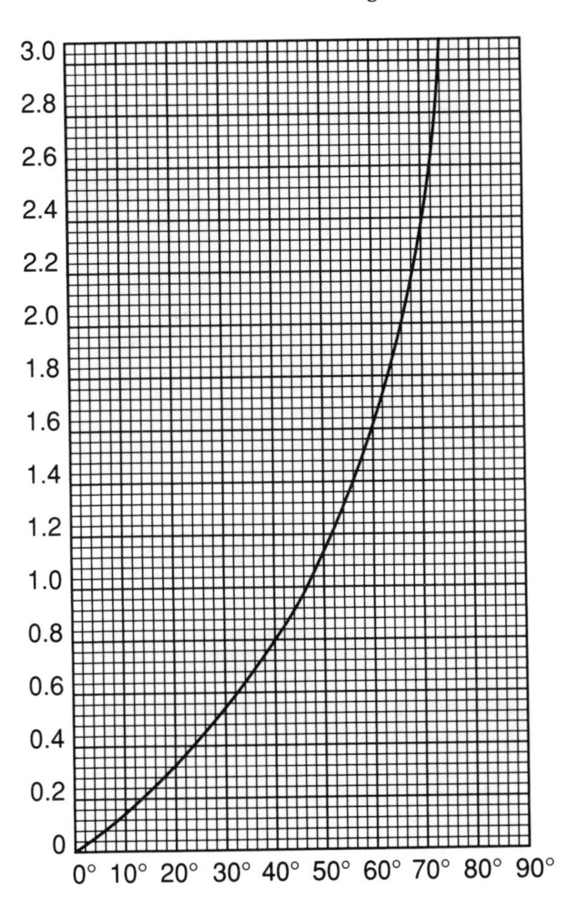


**Fig. 54.**
**Graph of *tan* θ.**

**Fig. 55.**

On your calculator the sequence of key presses should be:
260 ÷ 11 TAN = (or 260 × 11 TAN 1/X =)

*Example 3*: Find the value of $\frac{b - c}{b + c} \cot \frac{A}{2}$, when b = 25.6, c = 11.2, A = 57°.

Since $b = 25.6$
and $c = 11.2$

$$\therefore \quad b + c = 36.8$$
$$b - c = 14.4$$

and $$\frac{A}{2} = 57° \div 2 = 28.5°$$

Let $$x = \frac{b - c}{b + c} \cot \frac{A}{2}$$

Then $$x = \frac{14.4}{36.8} \cot 28.5°$$
$$= \frac{14.4}{36.8} \times \frac{1}{\tan 28.5°}$$
$$= 0.7207$$
$$\therefore \quad x = 0.7207.$$

On your calculator the sequence of key presses should be:
14.4 ÷ 36.8 ÷ 28.5 TAN = (or 14.4 ÷ 36.8 × 28.5 TAN 1/X =)

*Exercise 3*

**1** Find the following:

(1) cosec 35.4°
(2) cosec 59.75°
(3) sec 42.62°
(4) sec 53.08°
(5) cot 39.7°
(6) cot 70.57°

**2** Find the angle:

(1) When the cosecant is 1.1476
(2) When the secant is 2.3443
(3) When the cotangent is 0.3779

**3** The height of an isosceles triangle is 38 mm and each of the equal angles is 52°. Find the length of the equal sides.

**4** Construct a triangle with sides 5 cm, 12 cm and 13 cm in length. Find the cosecant, secant and tangent of each of the acute angles. Now use your calculator to find the angles.

**5** A chord of a circle is 3 m long and it subtends an angle of 63° at the centre. Find the radius of the circle.

**6** A man walks up a steep road the slope of which is 8°. What distance must he walk so as to rise 1 km?

**7** Find the values of:

(*a*) $\frac{8.72}{9.83} \sin 23°$

(*b*) $\cos A \sin B$ when $A = 40°$, $B = 35°$

**8** Find the values of:

(*a*) $\sin^2 \theta$ when $\theta = 28°$
(*b*) $2 \sec \theta \cot \theta$ when $\theta = 42°$

*Note* $\sin^2 \theta$ is the usual way of writing $(\sin \theta)^2$

**9** Find the values of:

(*a*) $\tan A \tan B$, when $A = 53°$, $B = 29°$

(*b*) $\frac{a \sin B}{b}$ when $a = 50$, $b = 27$, $B = 66°$

**10** Find the values of:

(*a*) $\sec^2 43°$ (*b*) $2 \cos^2 28°$

**11** Find the value of: $\sqrt{\frac{\sin 53.45°}{\tan 68.67°}}$

**12** Find the value of $\cos^2 \theta - \sin^2 \theta$

(1) When $\theta = 37.42°$ (2) When $\theta = 59°$

**13** If $\tan \frac{\theta}{2} = \sqrt{\frac{239 \times 25}{397 \times 133}}$ find $\theta$

**14** Find the value of $2 \sin \frac{A + B}{2} \cos \frac{A - B}{2}$ when $A = 57.23°$ and $B = 22.48°$

**15** If $\mu = \frac{\sin \theta}{\cot \alpha}$ find $\mu$ when $\theta = 10.42°$ and $\alpha = 28.12°$

**16** If $A = \frac{1}{2} ab \sin \theta$, find A when $a = 28.5$, $b = 46.7$ and $\theta = 56.28°$

# 61 Some applications of trigonometrical ratios

## Solution of right-angled triangles

By solving a right-angled triangle we mean, if certain sides or angles are given we require to find the remaining sides and angles.
Right-angled triangles can be solved:

(1) by using the appropriate trigonometrical ratios,
(2) By using the Theorem of Pythagoras (see Theorem 9, section 14).

We give a few examples.

***(a) Given the two sides which contain the right angle***
To solve this:

(1) The other angles can be found by the tangent ratios,
(2) The hypotenuse can be found by using secants and cosecants, or the Theorem of Pythagoras.

*Example 1*: Solve the right-angled triangle where the sides containing the right angle are 15.8 m and 8.9 m.

Fig. 56 illustrates the problem.

$$\text{To find C, } \tan C = \frac{8.9}{15.8} = 0.5633 = \tan 29.4°$$

$$\text{To find A, } \tan A = \frac{15.8}{8.9} = 1.7753 = \tan 60.6°$$

These should be checked by seeing if their sum is 90°.
To find AC.

(1) $AC = \sqrt{15.8^2 + 8.9^2} = 18.1$ m approx., or

(2) $$\frac{AC}{8.9} = \text{cosec } C$$

$$\therefore \quad AC = 8.9 \text{ cosec } C = 89 \div \sin C$$
$$= 18.13$$
$$\therefore \quad AC = 18.1 \text{ m (approx.)}.$$

***(b) Given one angle and the hypotenuse***
*Example 2*: Solve the right-angled triangle in which one angle is 27.72° and the hypotenuse is 6.85 cm.

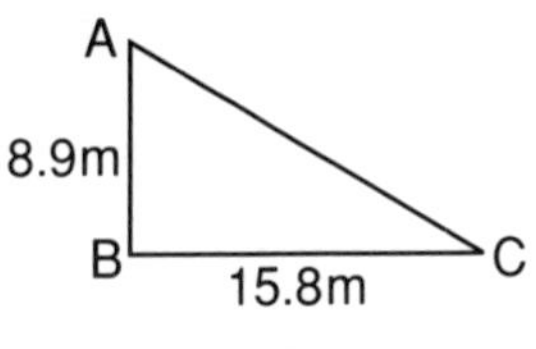


**Fig. 56.**

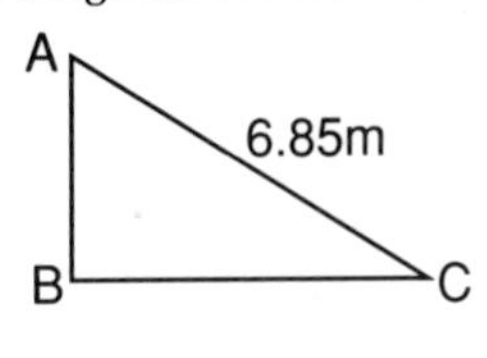


**Fig. 57.**

In Fig. 57

$$C = 27.72^\circ$$
$$\therefore \quad A = 90^\circ - C = 90 - 27.72^\circ$$
$$= 62.28^\circ$$

To find AB and BC

$$AB = AC \sin ACB$$
$$= 6.85 \times \sin 27.72^\circ$$
$$= 3.19 \text{ cm}$$
$$BC = AC \cos ACB$$
$$= 6.85 \times \cos 27.72^\circ$$
$$= 6.06 \text{ cm}$$

These examples will serve to indicate the methods to be adopted in other cases.

***(c) Special cases***

(1) The equilateral triangle

In Fig. 58 ABC is an equilateral triangle, AD is the perpendicular bisector of the base.

It also bisects $\angle CAB$ (Theorem 3, section 11).

$$\therefore \quad \angle DAB = 30^\circ$$
$$\text{and } \angle ABD = 60^\circ$$

Let each side of the $\triangle$ be a units of length.

$$\text{Then } DB = \frac{a}{2}$$

$$\therefore \quad AD = \sqrt{AB^2 - DB^2} \qquad \text{(Theorem 9)}$$
$$= \sqrt{a^2 - \frac{a^2}{4}}$$
$$= \sqrt{\frac{3a^2}{4}}$$
$$= a \times \frac{\sqrt{3}}{2}$$

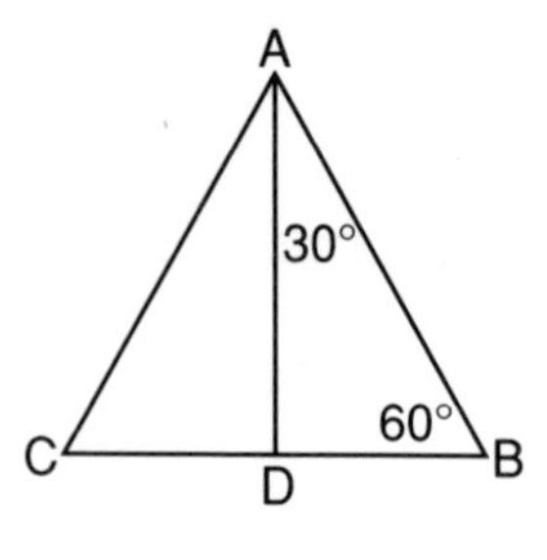


Fig. 58.

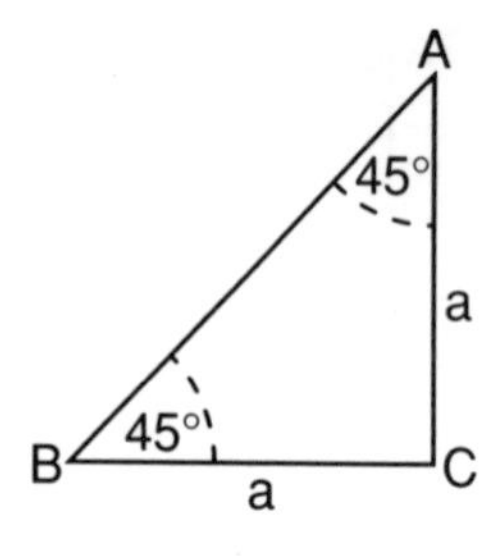


Fig. 59.

$$\therefore \quad \sin 60° = \frac{AD}{AB} = \frac{a \times \frac{\sqrt{3}}{2}}{a} = \frac{\sqrt{3}}{2}$$

$$\cos 60° = \frac{BD}{AB} = \frac{a}{2} \div a = \frac{1}{2}$$

$$\tan 60° = \frac{AD}{DB} = \frac{a\sqrt{3}}{2} \div \frac{a}{2} = \sqrt{3}$$

Similarly

$$\sin 30° = \frac{BD}{AB} = \frac{a}{2} \div a = \frac{1}{2}$$

$$\cos 30° = \frac{AD}{AB} = \frac{a\sqrt{3}}{2} \div a = \frac{\sqrt{3}}{2}$$

$$\tan 30° = \frac{DB}{AD} = \frac{a}{2} \div \frac{a\sqrt{3}}{2} = \frac{1}{\sqrt{3}}$$

*Note* The ratios for 30° can be found from those for 60° by using the results of sections 48 and 53.

(2) The right-angled isosceles triangle

Fig. 59 represents an isosceles triangle with AC = BC and $\angle ACB = 90°$.

Let each of the equal sides be a units of length.

Then $$AB^2 = AC^2 + BC^2 \qquad \text{(Theorem 9)}$$
$$= a^2 + a^2$$
$$= 2a^2$$
$$\therefore \quad AB = a\sqrt{2}$$

$$\therefore \quad \sin 45° = \frac{AC}{AB} = \frac{a}{a\sqrt{2}} = \frac{1}{\sqrt{2}}$$

$$\cos 45° = \frac{BC}{AB} = \frac{a}{a\sqrt{2}} = \frac{1}{\sqrt{2}}$$

$$\tan 45° = \frac{AC}{BC} = \frac{a}{a} = 1$$

It should be noted that △ABC represents half a square of which AB is the diagonal.

## 62 Slope and gradient

Fig. 60 represents a side view of the section of a path AC in which AB represents the horizontal level and BC the vertical rise.

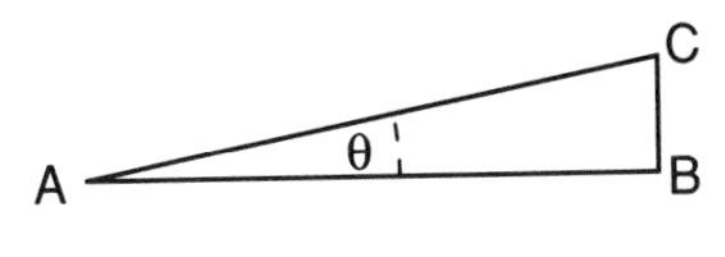


**Fig. 60.**

∠CAB, denoted by θ, is the angle between the plane of the path and the horizontal.

Then ∠CAB is called the **angle of slope** of the path or more briefly ∠CAB is the **slope** of the path.

Now $$\tan \theta = \frac{CB}{AB}$$

*This tangent is called the gradient of the path.*

Generally, if θ is the **slope** of a path, tan θ is the **gradient.**

A gradient is frequently given in the form 1 in 55, and in this form can be seen by the side of railways to denote the gradient of the rails. This means that the **tangent of the angle of slope** is $\frac{1}{55}$ .

When the angle of slope is very small, as happens in the case of a railway and most roads, it makes little practical difference if instead of the tangent $\left(\frac{CB}{AB}\right)$ we take $\frac{CB}{AC}$ i.e. the sine of the angle instead of the tangent. In practice also it is easier to measure AC, and the difference between this and AB is relatively small, provided the angle is small.

If you refer to the tables of tangents and sines you will see how small is the difference between them for small angles.

# 63 Projections

In Chapter 1, section 22, we referred to the projection of a straight line on a plane. We will now examine this further.

## Projection of a straight line on a fixed line

In Fig. 61, let PQ be a straight line of unlimited length, and AB another straight line which, when produced to meet PQ at O, makes an angle θ with it.

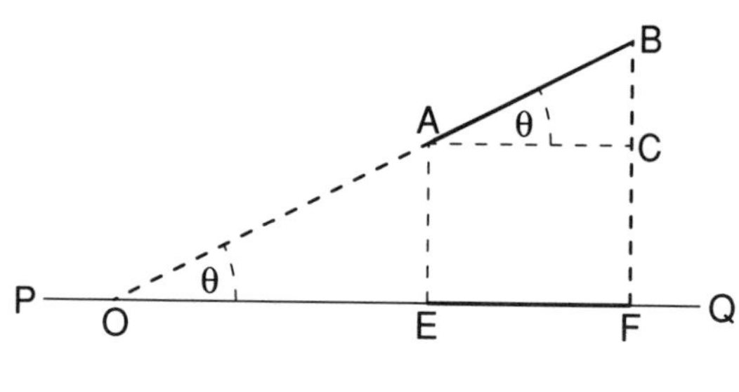


**Fig. 61.**

From A and B draw perpendiculars to meet PQ at E and F.
Draw AC parallel to EF.
EF is called the **projection** of AB on PQ (section 22).

Now $\quad \angle BAC = \angle BOF = \theta \quad$ (Theorem 2)
and $\quad EF = AC$
Also $\quad AC = AB \cos \theta \quad$ (section 47)
$\therefore \quad EF = AB \cos \theta$

∴ *If a straight line AB, produced if necessary, makes an angle θ with another straight line, the length of its projection on that straight line is AB cos θ.*

It should be noted in Fig. 61 that

$$BC = AB \sin \theta$$

From which it is evident that if we draw a straight line at right angles to PQ, the projection of AB upon such a straight line is AB sin θ.

*Exercise 4*

General questions on the trigonometrical ratios.

**1** In a right-angled triangle the two sides containing the right angle are 2.34 m and 1.64 m. Find the angles and the hypotenuse.

**2** In a triangle ABC, C being a right angle, AC is 122 cm, AB is 175 cm. Compute the angle B.

**3** In a triangle ABC, C = 90°. If A = 37.35° and c = 91.4, find a and b.

**4** ABC is a triangle, the angle C being a right angle. AC is 21.32 m, BC is 12.56 m. Find the angles A and B.

**5** In a triangle ABC, AD is the perpendicular on BC: AB is 3.25 cm, B is 55°, BC is 4.68 cm. Find the length of AD. Find also BD, DC and AC.

**6** ABC is a right-angled triangle, C being the right angle. If a = 378 mm and c = 543 mm, find A and b.

**7** A ladder 20 m long rests against a vertical wall. By means of trigonometrical tables find the inclination of the ladder to the horizontal when the foot of the ladder is:

(1) 7 m from the wall.
(2) 10 m from the wall.

**8** A ship starts from a point O and travels 18 km $h^{-1}$ in a direction 35° north of east. How far will it be north and east of O after an hour?

**9** A pendulum of length 20 cm swings on either side of the vertical through an angle of 15°. Through what height does the bob rise?

**10** If the side of an equilateral triangle is x m, find the altitude of the triangle. Hence find sin 60° and sin 30°.

**11** Two straight lines OX and OY are at right angles to one another. A straight line 3.5 cm long makes an angle of 42° with OX. Find the lengths of its projections on OX and OY.

**12** A man walking 1.5 km up the line of greatest slope of a hill rises 94 m. Find the gradient of the hill.

**13** A ship starts from a given point and sails 15.5 km in a direction 41.25° west of north. How far has it gone west and north respectively?

**14** A point P is 14.5 km north of Q and Q is 9 km west of R. Find the bearing of P from R and its distance from R.

# 4

# Relations between the Trigonometrical Ratios

## 64 The ratios

Since each of the trigonometrical ratios involves two of the three sides of a right-angled triangle, it is to be expected that definite relations exist between them. These relations are very important and will constantly be used in further work. The most important of them will be proved in this chapter.

$$\tan\theta\frac{\sin\theta}{\cos\theta}$$

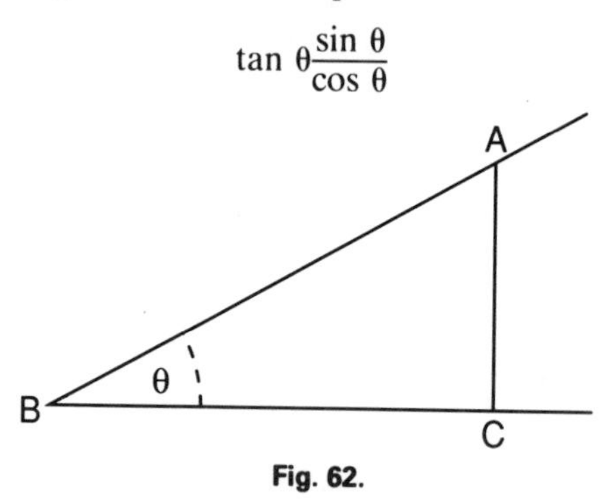


**Fig. 62.**

Let ABC (Fig. 62) be any acute angle (θ). From a point A on one arm draw AC perpendicular to the other arm.

Then $$\sin\theta = \frac{AC}{AB}$$

and $$\cos\theta = \frac{BC}{AB}$$

$$\therefore \quad \frac{\sin\theta}{\cos\theta} = \frac{AC}{AB} \div \frac{BC}{AB}$$

$$= \frac{AC}{AB} \div \frac{AB}{BC}$$

$$= \frac{AC}{BC}$$

$$= \tan\theta$$

$$\therefore \quad \frac{\sin\theta}{\cos\theta} = \tan\theta \qquad (1)$$

Similarly we may prove that $\cot\theta = \dfrac{\cos\theta}{\sin\theta}$

## 65 $\sin^2\theta + \cos^2\theta = 1$

From Fig. 62

$AC^2 + BC^2 = AB^2$ (Theorem of Pythagoras, section 14)

Dividing throughout by $AB^2$

we get
$$\frac{AC^2}{AB^2} + \frac{BC^2}{AB^2} = 1$$

$$\therefore \quad (\sin\theta)^2 + (\cos\theta)^2 = 1$$

or as usually written

$$\sin^2\theta + \cos^2\theta = 1 \qquad (2)$$

This very important result may be transformed and used to find either of the ratios when the other is given.

Thus
$$\sin^2\theta = 1 - \cos^2\theta$$
$$\therefore \quad \sin\theta = \sqrt{1 - \cos^2\theta}$$

Similarly
$$\cos\theta = \sqrt{1 - \sin^2\theta}$$

Combining formulae (1) and (2)

$$\tan\theta = \frac{\sin\theta}{\cos\theta}$$

becomes
$$\tan\theta = \frac{\sin\theta}{\sqrt{1 - \sin^2\theta}}$$

This form expresses the **tangent** in terms of the **sine** only. It may similarly be expressed in terms of the cosine

thus $$\tan \theta = \frac{\sqrt{1 - \cos^2 \theta}}{\cos \theta}$$

**66**

$$1 + \tan^2 \theta = \sec^2 \theta$$
$$1 + \cot^2 \theta = \operatorname{cosec}^2 \theta$$

Using the formula $\sin^2 \theta + \cos^2 \theta = 1$
and dividing throughout by $\cos^2 \theta$

we get $$\frac{\sin^2 \theta}{\cos^2 \theta} + 1 = \frac{1}{\cos^2 \theta}$$

$$\therefore \quad \tan^2 \theta + 1 = \sec^2 \theta$$

Again, dividing throughout by $\sin^2 \theta$

we get $$1 + \frac{\cos^2 \theta}{\sin^2 \theta} = \frac{1}{\sin^2 \theta}$$

$$\therefore \quad 1 + \cot^2 \theta = \operatorname{cosec}^2 \theta$$

We may also write these formulae in the forms

$$\tan^2 \theta = \sec^2 \theta - 1$$
and $$\cot^2 \theta = \operatorname{cosec}^2 \theta - 1$$

Using these forms we can change tangents into secants and cotangents into cosecants and vice versa when it is necessary in a given problem.

### *Exercise 5*

**1** Find $\tan \theta$ when $\sin \theta = 0.5736$ and $\cos \theta = 0.8192$.
**2** If $\sin \theta = \frac{3}{5}$, find $\cos \theta$ and $\tan \theta$.
**3** Find $\sin \theta$ when $\cos \theta = 0.47$.
**4** Find $\sec \theta$ when $\tan \theta = 1.2799$.
**5** If $\sec \theta = 1.2062$ find $\tan \theta$, $\cos \theta$ and $\sin \theta$.
**6** Find $\operatorname{cosec} \theta$ when $\cot \theta = 0.5774$.
**7** If $\cot \theta = 1.63$, find $\operatorname{cosec} \theta$, $\sin \theta$ and $\cos \theta$.
**8** If $\tan \theta = t$, find expressions for $\sec \theta$, $\cos \theta$ and $\sin \theta$ in terms of t.
**9** If $\cos \alpha = 0.4695$, find $\sin \alpha$ and $\tan \alpha$.
**10** Prove that $\tan \theta + \cot \theta = \sec \theta \operatorname{cosec} \theta$.

# 5

# Trigonometrical Ratios of Angles in the Second Quadrant

## 67 Extending the ratios

In chapter 3 we dealt with the trigonometrical ratios of **acute** anges, or angles in the first quadrant. In chapter 1, section 5, when considering the meaning of an angle as being formed by the rotation of a straight line from a fixed position, we saw that there was no limit to the amount of rotation, and consequently that angles could be of any magnitude.

We must now consider the extension of trigonometrical ratios to angles greater than a right angle. At the present, however, we shall not examine the general question of angles of any magnitude, but confine ourselves to **obtuse** angles, or angles in the second quadrant, as these are necessary in many practical applications of trigonometry.

## 68 Positive and negative lines

Before proceeding to deal with the trigonometrical ratios of obtuse angles it is necessary to consider the methods by which we distinguish between measurements made on a straight line in opposite directions. These will be familiar to those who have studied co-ordinates and graphs. It is desirable, however, to revise the principles involved before applying them to trigonometry.

Let Fig. 63 represent a straight road XOX′.

If a man now travels 4 miles from O to P in the direction $\overrightarrow{OX}$ and then turns and travels 6 miles in the opposite direction to P′,

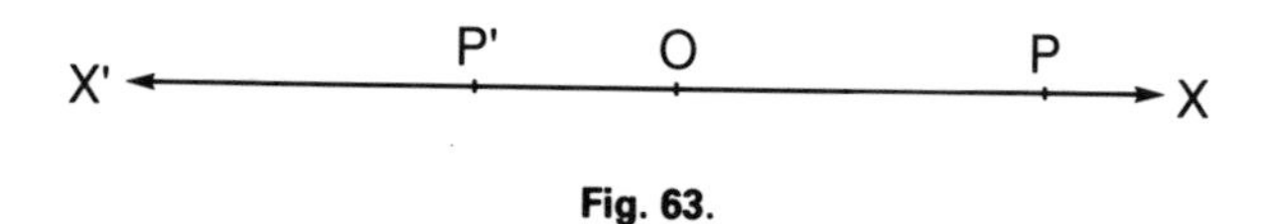


Fig. 63.

the net result is that he has travelled (4–6) miles, i.e. −2 miles from O. The significance of the negative sign is that the man is now 2 miles in the opposite direction from that in which he started.

In such a way as this we arrive at the convention by which we agree to use + and − signs to indicate opposite directions.

If we now consider two straight lines at right angles to one another, as X′OX, Y′OY, in Fig. 64, such as are used for co-ordinates and graphs, we can extend to these the conventions used for one straight line as indicated above. The lines OX, OY are called the **axes of co-ordinates**. OX measures the x-co-ordinate, called the **abscissa**, and OY measures the y-co-ordinate, called the **ordinate**. Any point P (Fig. 64), has a **pair** of co-ordinates (x, y). Each pair determines a unique point.

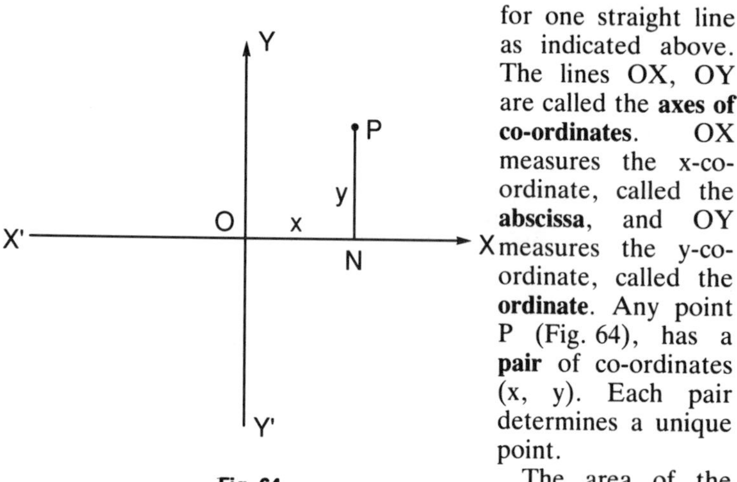


Fig. 64.

The area of the diagram, Fig. 65, is considered to be divided into four quadrants as shown. Values of x measured to the right are +ve, and to the left are −ve. Values of y measured upwards are +ve, and downwards are −ve. This is a universally accepted convention.

$P_1$ lies in the first quadrant and $N_1$ is the foot of the perpendicular from $P_1$ to OX. $\overrightarrow{ON_1}$ is in the direction of $\overrightarrow{OX}$ and is +ve; $N_1P$ is in the direction of $\overrightarrow{OY}$ and is +ve. Thus the co-ordinates of any point $P_1$ in the first quadrant are (+, +).

$P_2$ lies in the second quadrant and $N_2$ is the foot of the

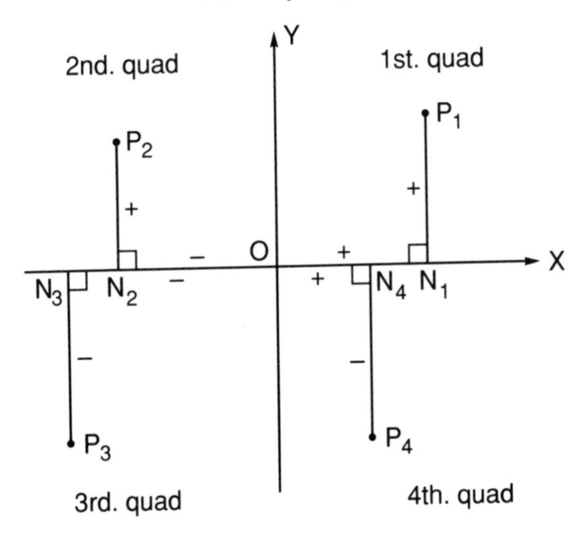


**Fig. 65.**

perpendicular from $P_2$ to OX. $\overrightarrow{ON_2}$ is in the direction of $\overrightarrow{XO}$ and is − ve; $\overrightarrow{N_2P_2}$ is in the direction of $\overrightarrow{OY}$ and is + ve. Thus the co-ordinates of any point $P_2$ in the second quadrant are (−, +).

Similarly the co-ordinates of $P_3$ in the third quadrant are (−, −) and of $P_4$ in the fourth quadrant are (+, −).

At present we shall content ourselves with considering points in the first two quadrants. The general problem for all four quadrants is discussed later (chapter 11).

## 69 Direction of rotation of angle

The **direction** in which the rotating line turns must be taken into account when considering the angle itself.

Thus in Fig. 66 the angle AOB may be formed by rotation in an anti-clockwise direction or by rotation in a clockwise direction.

*By convention an anti-clockwise rotation is positive and a clockwise rotation is negative.*

Negative angles will be considered further in chapter 11. In the meantime, we shall use positive angles formed by anti-clockwise rotation.

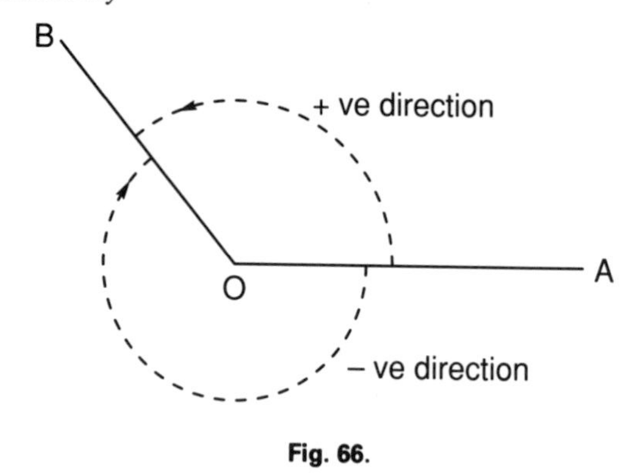


**Fig. 66.**

## 70 The sign convention for the hypotenuse

Consider a point A in the first quadrant. Draw AD perpendicular to X′OX meeting it at D (Fig. 67).

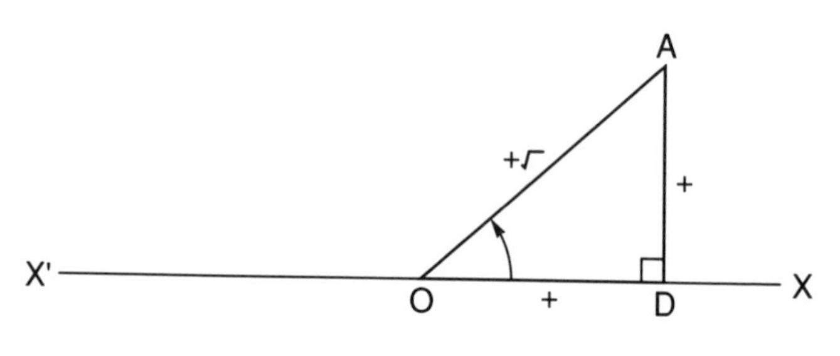


**Fig. 67.**

OD is +ve and DA is +ve. The angle XOA = angle DOA, which is acute.

Also $\quad OA^2 = OD^2 + DA^2$
$= (+\text{ve})^2 + (+\text{ve})^2 = +\text{ve quantity}$
$= a^2$ (say where a is +ve)

Now the equation $OA^2 = a^2$ has two roots OA = a or OA = −a, so we must decide on a sign convention. *We take OA as the +ve root.*

Now consider a point B in the second quadrant. Draw BE perpendicular to X′OX meeting it at E (Fig. 68).

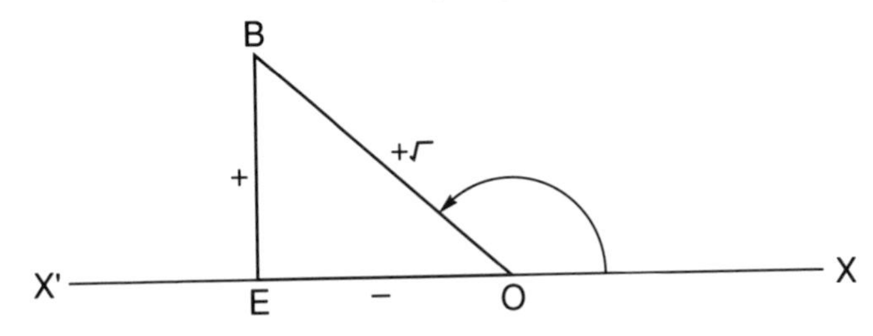


**Fig. 68.**

OE is −ve and EB is +ve. The angle XOB (= 180° − angle EOB) is obtuse.

Also $$\begin{aligned} OB^2 &= OE^2 + EB^2 \\ &= (-\text{ve})^2 + (+\text{ve})^2 \\ &= (+\text{ve}) + (+\text{ve}) = +\text{ve quantity} \end{aligned}$$

We have already decided on a sign convention for the root, so OB is +ve.

Now the sides required to give the ratios of ∠XOB are the same as those needed for its supplement ∠EOB. The only change which may have taken place is in the sign prefixed to the length of a side. OD (+ve in Fig. 67) has become OE (−ve in Fig. 68).

Thus we have the following rules:

| RATIO | ACUTE ANGLE | OBTUSE ANGLE |
|---|---|---|
| SIN | + | + |
| COS | + | − |
| TAN | + | − |

**Fig. 69.**

We see this at once by combining Fig. 67 and Fig. 68 into Fig. 70.

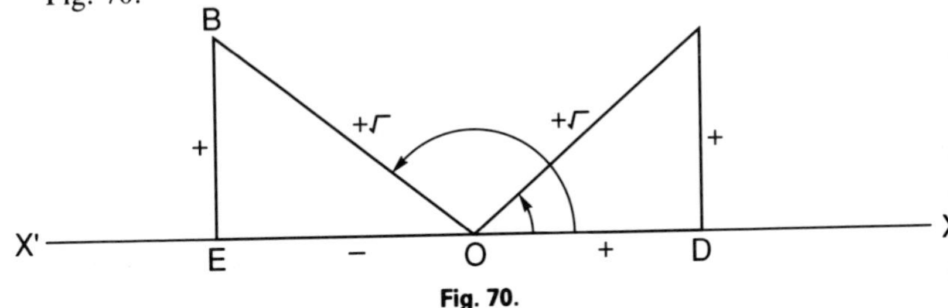


**Fig. 70.**

$$\sin XOA = \frac{DA}{OA} = \frac{+}{+\sqrt{}} = + \text{ (see } \textit{Note}\text{)}$$

$$\sin XOB = \frac{EB}{OB} = \frac{+}{+\sqrt{}} = +$$

$$\cos XOA = \frac{OD}{OA} = \frac{+}{+\sqrt{}} = +$$

$$\cos XOB = \frac{OE}{OB} = \frac{-}{+\sqrt{}} = -$$

$$\tan XOA = \frac{DA}{OD} = \frac{+}{+} = +$$

$$\tan XOB = \frac{EB}{OE} = \frac{+}{-} = -$$

*Note* We use here the abbreviations + and − to stand for **a positive quantity** and **a negative quantity** respectively.

In addition, by making $\triangle OBE \equiv \triangle AOD$ in Fig. 70 and using the rules we see that

**sine of an angle = sine of its supplement**
**cosine of an angle = − cosine of its supplement**
**tangent of an angle = − tangent of its supplement**

These results may alternatively be expressed thus:

$$\sin\theta = \sin(180° - \theta)$$
$$\cos\theta = -\cos(180° - \theta)$$
$$\tan\theta = -\tan(180° - \theta)$$

e.g.

$$\left.\begin{array}{l} \sin 100° = \sin 80° \\ \cos 117° = -\cos 63° \\ \tan 147° = -\tan 33° \end{array}\right\}$$

The reciprocal ratios, cosecant, secant and cotangent will have the same signs as the ratios from which they are derived.

$\therefore$ **cosecant** has same sign as **sine**
**secant** has same sign as **cosine**
**cotangent** has same sign as **tangent**

e.g.

$$\text{cosec } 108° = \text{cosec } 72°$$
$$\sec 121° = -\sec 59°$$
$$\cot 154° = -\cot 36°$$

## 71 To find the ratios of angles in the second quadrant from the tables

As will have been seen, the tables of trigonometrical ratios give the ratios of angles in the first quadrant only. But each of these is supplementary to an angle in the second quadrant. Consequently if a ratio of an angle in the second quadrant is required, we find its supplement which is an angle in the first quadrant, and then, by using the relations between the two angles as shown in the previous paragraph we can write down the required ratio from the tables.

*Example 1*: Find from the tables sin 137° and cos 137°.

We first find the supplement of 137° which is

$$180° - 137° = 43°.$$

$\therefore$ by section 70 $\sin 137° = \sin 43°$

From the tables $\sin 43° = 0.6820$

$$\therefore \quad \sin 137° = 0.6820$$

Again

$$\cos\theta = -\cos(180° - \theta)$$
$$\therefore \quad \cos 137° = -\cos(180° - 137°)$$
$$= -\cos 43°$$
$$= -0.7314$$

*Example 2*: Find the values of tan 162° and sec 162°.

From the above

$$\tan\theta = -\tan(180° - \theta)$$
$$\therefore \quad \tan 162° = -\tan(180° - 162°)$$
$$= -\tan 18°$$
$$= -0.3249$$

Also

$$\sec\theta = -\sec(180° - \theta)$$
$$\therefore \quad \sec 162° = -\sec(180° - 162°)$$
$$= -\sec 18°$$
$$= -1.0515$$

## 72 Ratios for 180°

These can be found either by using the same arguments as were employed in the cases of 0° and 90° or by applying the above relation between an angle and its supplement.

From these we conclude

$$\mathbf{\sin 180° = 0}$$
$$\mathbf{\cos 180° = -1}$$
$$\mathbf{\tan 180° = 0}$$

## 73 To find an angle when a ratio is given

When this converse problem has to be solved in cases where the angle may be in the second quadrant, difficulties arise which did not occur when dealing with angles in the first quadrant only. The following examples will illustrate these.

*Example 1*: Find the angle whose cosine is $-0.5577$.

The negative sign for a cosine shows that the angle is in the second quadrant, since $\cos\theta = -\cos(180° - \theta)$.

From the tables we find that

$$\cos 56.1° = +0.5577$$

$\therefore$ the angle required is the supplement of this

$$\text{i.e.}\quad 180° - 56.1° = 123.9°$$

*Example 2*: Find the angles whose sine is $+0.9483$.

We know that since an angle and its supplement have the same sine, there are two angles with the sine $+0.9483$, and they are supplementary.

From the tables

$$\sin 71.5° = +0.9483$$

$\therefore$ Since

$$\sin\theta = \sin(180° - \theta)$$

$$\therefore \sin 71.5° = \sin(180° - 71.5°) = \sin 108.5°$$

There are therefore two answers, 71.5° and 108.5° and there are always two angles having a given sine, one in the first and one in the second quadrant. Which of these is the angle required when solving some problem must be determined by the special conditions of the problem.

*Example 3*: Find the angle whose tangent is $-1.3764$.

Since the tangent is negative, the angle required must lie in the second quadrant.

From the tables

$$\tan 54° = +1.3764$$

and since

$$\tan\theta = -\tan(180° - \theta)$$

$$\therefore -1.3764 = \tan(180° - 54°) = \tan 126°$$

## 74 Inverse notation

The sign $tan^{-1} - 1.3674$ is employed to signify *the angle whose tangent is* $-1.3674$.

And, in general

$sin^{-1}x$ means *the angle whose sine is x*
$cos^{-1}x$ means *the angle whose cosine is x*

Three points should be noted.

(1) $\sin^{-1}x$ stands for an **angle**: thus $\sin^{-1}\frac{1}{2} = 30°$.
(2) The '$-1$' is not an index, but merely a sign to denote inverse notation.
(3) $(\sin x)^{-1}$ is not used, because by section 31 it would mean the reciprocal of sin x and this is cosec x.

## 75 Ratios of some important angles

We are now able to tabulate the values of the sine, cosine and tangents of certain angles between 0° and 180°. The table will also state in a convenient form the ratios of a few important angles. They are worth remembering.

| | 0° | 30° | 45° | 60° | 90° | 120° | 135° | 150° | 180° |
|---|---|---|---|---|---|---|---|---|---|
| | Increasing and Positive | | | | | Decreasing and Positive | | | |
| Sine | 0 | $\frac{1}{2}$ | $\frac{1}{\sqrt{2}}$ | $\frac{\sqrt{3}}{2}$ | 1 | $\frac{\sqrt{3}}{2}$ | $\frac{1}{\sqrt{2}}$ | $\frac{1}{2}$ | 0 |
| | Decreasing and Positive | | | | | Decreasing and Negative | | | |
| Cosine | 1 | $\frac{\sqrt{3}}{2}$ | $\frac{1}{\sqrt{2}}$ | $\frac{1}{2}$ | 0 | $-\frac{1}{2}$ | $-\frac{1}{\sqrt{2}}$ | $-\frac{\sqrt{3}}{2}$ | $-1$ |
| | Increasing and Positive | | | | | Increasing and Negative | | | |
| Tangent | 0 | $\frac{1}{\sqrt{3}}$ | 1 | $\sqrt{3}$ | $\infty$ | $-\sqrt{3}$ | $-1$ | $-\frac{1}{\sqrt{3}}$ | 0 |

## 76 Graphs of sine, cosine and tangent between 0° and 180°

The changes in the ratios of angles in the first and second quadrants are made clear by drawing their graphs. This may be

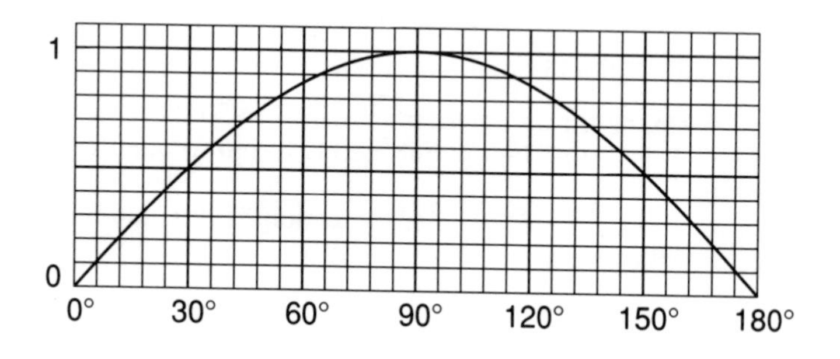


**Fig. 71. Sin θ.**

done by using the values given in the above table or, more accurately, by taking values from the tables.

An inspection of these graphs will illustrate the results reached in section 73 (second example).

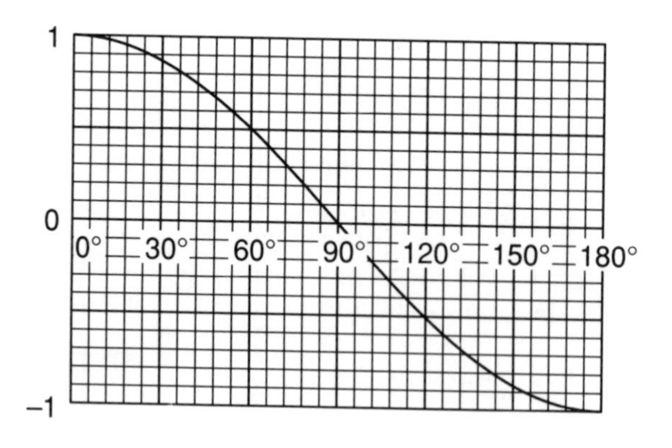


**Fig. 72. Cos θ.**

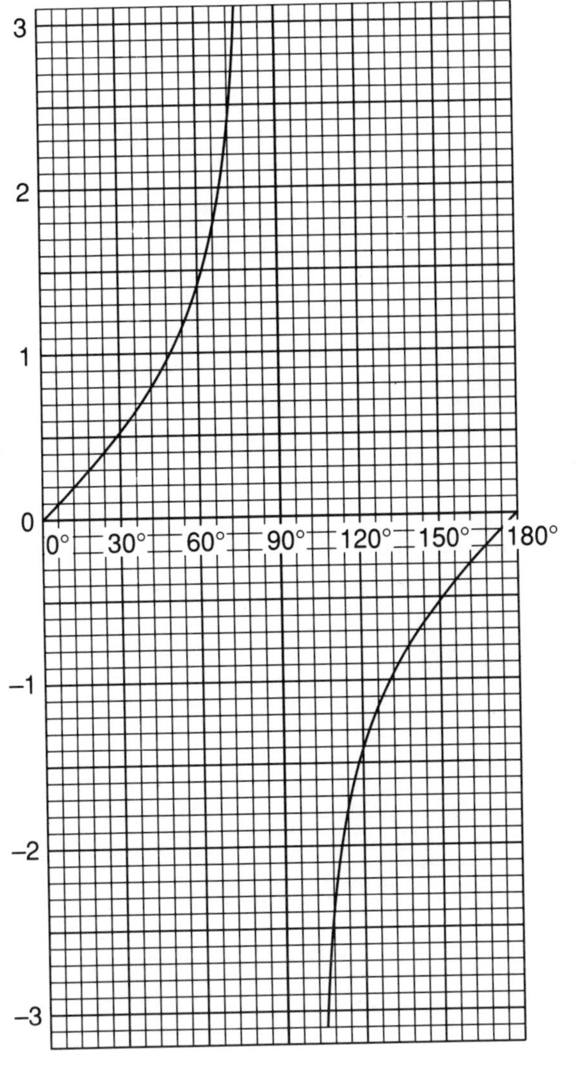


**Fig. 73. Tan θ.**

It is evident from Fig. 71, that there are two angles, one in each quadrant with a given sine.

From Figs. 72 and 73, it will be seen that there is only one angle between 0° and 180° corresponding to a given cosine or tangent.

## *Exercise 6*

**1** Write down from the tables the sines, cosines and tangents of the following angles:

(*a*) 102° (*b*) 149.55° (*c*) 109.47°
(*d*) 145.27° (*e*) 154° 36°

Check your results on your calculator.

**2** Find two values of $\theta$ between 0° and 180° when:

(*a*) $\sin\theta = 0.6508$ (*b*) $\sin\theta = 0.9126$
(*c*) $\sin\theta = 0.3469$ (*d*) $\sin\theta = 0.7122$

**3** Find the angles between 0° and 180° whose cosines are:

(*a*) $-0.4540$ (*b*) $-0.8131$ (*c*) $-0.1788$
(*d*) $-0.9354$ (*e*) $-0.7917$ (*f*) $-0.9154$

**4** Find $\theta$ between 0° and 180° when:

(*a*) $\tan\theta = -0.5543$ (*b*) $\tan\theta = -1.4938$
(*c*) $\tan\theta = 2.4383$ (*d*) $\tan\theta = -1.7603$
(*e*) $\tan\theta = -0.7142$ (*f*) $\tan\theta = -1.1757$

**5** Find the values of:

(*a*) cosec 154° (*b*) sec 162.5°
(*c*) cot 163.2°

**6** Find $\theta$ between 0° and 180° when:

(*a*) $\sec\theta = -1.6514$ (*b*) $\sec\theta = -2.1301$
(*c*) $\operatorname{cosec}\theta = 1.7305$ (*d*) $\operatorname{cosec}\theta = 2.4586$
(*e*) $\cot\theta = -1.6643$ (*f*) $\cot\theta = -0.3819$

**7** Find the value of $\frac{\tan A}{\sec B}$ when $A = 150°$, $B = 163.28°$.

**8** Find the angles between 0° and 180° when:

(*a*) $\sin^{-1} 0.9336$ (*b*) $\cos^{-1} 0.4226$
(*c*) $\tan^{-1} 1.3764$ (*d*) $\cos^{-1} - 0.3907$

# 6

# Trigonometrical Ratios of Compound Angles

**77** We often need to use the trigonometrical ratios of the sum or difference of two angles. If A and B are any two angles, (A + B) and (A − B) are usually called compound angles, and it is convenient to be able to express their trigonometrical ratios in terms of the ratios of A and B.

Beware of thinking that sin (A + B) is equal to (sin A + sin B). You can test this by taking the values of sin A, sin B, and sin (A + B) for some particular values of A and B from the tables and comparing them.

**78** We will first show that:

$$\sin (A + B) = \sin A \cos B + \cos A \sin B$$

and

$$\cos (A + B) = \cos A \cos B - \sin A \sin B$$

To simplify the proof at this stage we will assume that A, B, and (A + B) are all acute angles.

You may find it useful to make your own diagram step by step with the following construction.

### Construction

Let a straight line rotating from a position on a fixed line OX trace out (1) the angle XOY, equal to A and YOZ equal to B (Fig. 74).

Then $\angle XOZ = (A + B)$

In OZ take any point P.
Draw PQ perpendicular to OX and PM perpendicular to OY.
From M draw MN perpendicular to OX and MR parallel to OX.

Then $\quad MR = QN$

*Proof*

$$\angle RPM = 90^\circ - \angle PMR$$
$$= RMO$$

But $\quad \angle RMO = \angle MOZ \qquad$ (Theorem 2, section 9)

$$= A$$
$$\therefore \quad \angle RPM = A$$

Again
$$\sin(A + B) = \sin XOZ$$
$$= \frac{PQ}{OP}$$
$$= \frac{RQ + PR}{OP}$$
$$= \frac{RQ}{OP} + \frac{PR}{OP}$$
$$= \frac{MN}{OP} + \frac{PR}{OP}$$
$$= \left(\frac{MN}{OM} \times \frac{OM}{OP}\right) + \left(\frac{PR}{PM} \times \frac{PM}{OP}\right)$$
$$= \sin A \cos B + \cos A \sin B$$

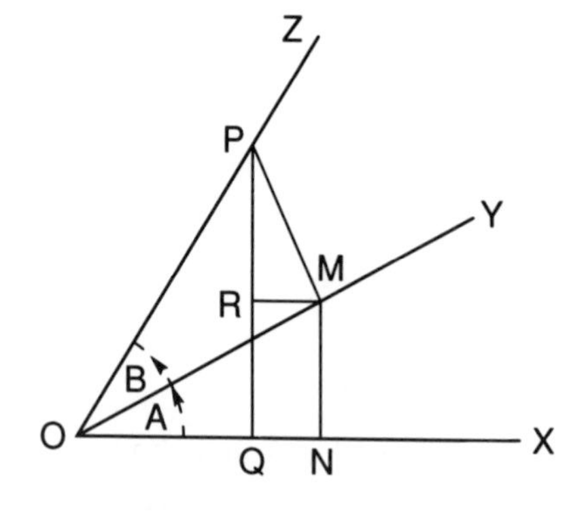


**Fig. 74.**

*Note* the device of introducing $\frac{OM}{MO}$ and $\frac{PM}{PM}$, each of which is unity, into the last line but one.

Again

$$\begin{aligned}\cos(A + B) &= \cos XOZ \\ &= \frac{OQ}{OP} \\ &= \frac{ON - NQ}{OP} \\ &= \frac{ON}{OP} - \frac{NQ}{OP} \\ &= \frac{ON}{OP} - \frac{RM}{OP} \\ &= \left(\frac{ON}{OM} \times \frac{OM}{OP}\right) - \left(\frac{RM}{PM} \times \frac{PM}{OP}\right) \\ &= \cos A \cos B - \sin A \sin B\end{aligned}$$

**79** We will next prove the corresponding formulae for $(A - B)$, viz.:

$$\sin(A - B) = \sin A \cos B - \cos A \sin B$$
$$\cos(A - B) = \cos A \cos B + \sin A \sin B$$

## Construction

Let a straight line rotating from a fixed position on OX describe an angle XOY, equal to A, and then, rotating in an opposite direction, describe an angle YOZ, equal to B (Fig. 75).

Then $$XOZ = A - B$$

Take a point P on OZ.
Draw PQ perpendicular to OX and PM perpendicuar to OY.
From M draw MN perpendicular to OX and MR parallel to OX to meet PQ **produced** in R.

*Proof*

$$\begin{aligned}\angle RPM &= 90° - \angle PMR \\ &= \angle RMY && \text{(since PM is perp. to OY)} \\ &= \angle YOX && \text{(Theorem 2, section 9)} \\ &= A\end{aligned}$$

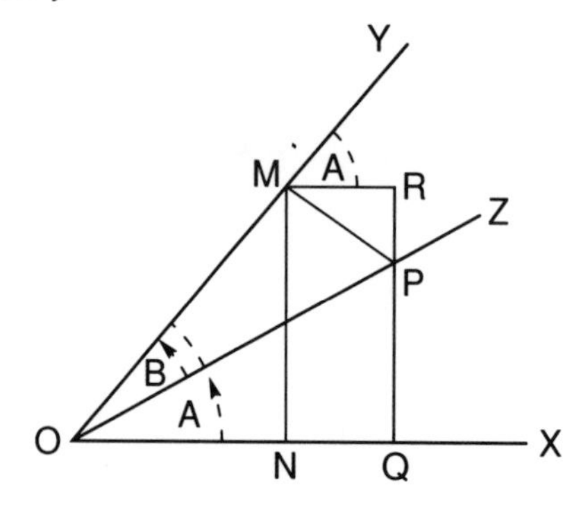


**Fig. 75.**

Now

$$\sin (A - B) = \sin XOZ$$

$$= \frac{PQ}{OP}$$

$$= \frac{RQ - RP}{OP}$$

$$= \frac{RQ}{OP} - \frac{RP}{OP}$$

$$= \frac{MN}{OP} - \frac{RP}{OP}$$

$$= \left(\frac{MN}{OM} \times \frac{OM}{OP}\right) - \left(\frac{RP}{PM} \times \frac{PM}{OP}\right)$$

$$= \sin A \cos B - \cos A \sin B$$

Again

$$\cos (A - B) = \cos XOZ$$

$$= \frac{OQ}{OP}$$

$$= \frac{ON + QN}{OP}$$

$$= \frac{ON}{OP} + \frac{QN}{OP}$$

$$= \frac{ON}{OP} + \frac{RM}{OP}$$

$$= \left(\frac{ON}{OM} \times \frac{OM}{OP}\right) + \left(\frac{RM}{PM} \times \frac{PM}{OP}\right)$$

$$= \cos A \cos B + \sin A \sin B$$

**80** These formulae have been proved for acute angles only, but they can be shown to be true for angles of any size. They are of great importance. We collect them for reference:

$$\sin(A + B) = \sin A \cos B + \cos A \sin B \quad (1)$$
$$\cos(A + B) = \cos A \cos B - \sin A \sin B \quad (2)$$
$$\sin(A - B) = \sin A \cos B - \cos A \sin B \quad (3)$$
$$\cos(A - B) = \cos A \cos B + \sin A \sin B \quad (4)$$

**81** From the above we may find similar formulae for $\tan(A + B)$ and $\tan(A - B)$ as follows:

$$\tan(A + B) = \frac{\sin(A + B)}{\cos(A + B)}$$

$$= \frac{\sin A \cos B + \cos A \sin B}{\cos A \cos B - \sin A \sin B}$$

Dividing numerator and denominator by $\cos A \cos B$

we get $$\tan(A + B) = \frac{\dfrac{\sin A \cos B}{\cos A \cos B} + \dfrac{\cos A \sin B}{\cos A \cos B}}{\dfrac{\cos A \cos B}{\cos A \cos B} - \dfrac{\sin A \sin B}{\cos A \cos B}}$$

$$= \frac{\dfrac{\sin A}{\cos A} + \dfrac{\sin B}{\cos B}}{1 - \dfrac{\sin A}{\cos A} \cdot \dfrac{\sin B}{\cos B}}$$

$$\therefore \quad \tan(A + B) = \frac{\tan A + \tan B}{1 - \tan A \tan B}$$

Similarly we may show

$$\tan(A - B) = \frac{\tan A - \tan B}{1 + \tan A \tan B}$$

with similar formulae for cotangents.

## 82 Worked Examples

*Example 1*: Using the values of the sines and cosines of 30° and 45° as given in the table in section 75, find sin 75°.

Using

$$\sin (A + B) = \sin A \cos B + \cos A \sin B$$

and substituting

$$A = 45°, B = 30°$$

we have
$$\sin 75° = \sin 45° \cos 30° + \cos 45° \sin 30°$$
$$= \left(\frac{1}{\sqrt{2}} \times \frac{\sqrt{3}}{2}\right) + \left(\frac{1}{\sqrt{2}} \times \frac{1}{2}\right)$$
$$= \frac{\sqrt{3}}{2\sqrt{2}} + \frac{1}{2\sqrt{2}}$$
$$= \frac{\sqrt{3} + 1}{2\sqrt{2}}$$

*Example 2*: If $\cos \alpha = 0.6$ and $\cos \beta = 0.8$, find the values of $\sin (\alpha + \beta)$ and $\cos (\alpha + \beta)$, without using the tables.

We must first find $\sin \alpha$ and $\sin \beta$. For these we use the results given in section 65.

$$\sin \alpha = \sqrt{1 - \cos^2 \alpha}$$

Substituting the given value of $\cos \alpha$

$$\begin{aligned} \sin \alpha &= \sqrt{1 - (0.6)^2} \\ &= \sqrt{1 - 0.36} \\ &= \sqrt{0.64} \\ &= \mathbf{0.8} \end{aligned}$$

Similarly we find $\sin \beta = 0.6$.

Using $\sin (A + B) = \sin A \cos B + \cos A \sin B$ and substituting we have

$$\begin{aligned} \sin (\alpha + \beta) &= (0.8 \times 0.8) + (0.6 \times 0.6) \\ &= 0.64 + 0.36 \\ &= 1 \end{aligned}$$

Also
$$\begin{aligned} \cos (\alpha + \beta) &= \cos \alpha \cos \beta - \sin \alpha \sin \beta \\ &= (0.6 \times 0.8) - (0.8 \times 0.6) \\ &= 0 \end{aligned}$$

Obviously $\alpha + \beta = 90°$, since $\cos 90° = 0$.

$\therefore$ $\alpha$ and $\beta$ are complementary.

*Exercise 7*

**1** If $\cos A = 0.2$ and $\cos B = 0.5$, find the values of $\sin (A + B)$ and $\cos (A - B)$.

**2** Use the ratios of 45° and 30° from the table in section 75 to find the values of sin 15° and cos 75°.

**3** By using the formula for $\sin (A - B)$ prove that:

$$\sin (90° - \theta) = \cos \theta.$$

**4** By means of the formulae of section 80, find $\sin (A - B)$ when $\sin B = 0.23$ and $\cos A = 0.309$.

**5** Find $\sin (A + B)$ and $\tan (A + B)$ when $\sin A = 0.71$ and $\cos B = 0.32$.

**6** Use the formula of $\tan (A + B)$ to find tan 75°.

**7** Find $\tan (A + B)$ and $\tan (A - B)$ when $\tan A = 1.2$ and $\tan B = 0.4$.

**8** By using the formula for $\tan (A - B)$ prove that

$$\tan (180° - A) = -\tan A.$$

**9** Find the values of:

(1) $\sin 52° \cos 18° - \cos 52° \sin 18°$.
(2) $\cos 73° \cos 12° + \sin 73° \sin 12°$.

**10** Find the values of: (*a*) $\dfrac{\tan 52° + \tan 16°}{1 - \tan 52° \tan 16°}$

(*b*) $\dfrac{\tan 64° - \tan 25°}{1 + \tan 64° \tan 25°}$

**11** Prove that $\sin (\theta + 45°) = \dfrac{1}{\sqrt{2}}(\sin \theta + \cos \theta)$

**12** Prove that $\tan (\theta + 45°) = \dfrac{\tan \theta + 1}{1 - \tan \theta}$

## 83 Multiple and sub-multiple angle formulae

From the preceding formulae we may deduce others of great practical importance.

From section 78 $\sin (A + B) = \sin A \cos B + \cos A \sin B$.

There have been no limitations of the angles.

∴ let $B = A.$

Substituting

$$\sin 2A = \sin A \cos A + \cos A \sin A$$

or

$$\sin 2A = 2 \sin A \cos A \quad (1)$$

If 2A is replaced by θ

then

$$\sin \theta = 2 \sin \frac{\theta}{2} \cos \frac{\theta}{2} \quad (2)$$

We may use whichever of these formulae is more convenient in a given problem.

Again $\cos (A + B) = \cos A \cos B - \sin A \sin B$

Let $B = A,$

then

$$\cos 2A = \cos^2 A - \sin^2 A \quad (4)$$

This may be transformed into formulae giving cos A or $\sin^2 A$ in terms of $^2$ A.

Since $\sin^2 A + \cos^2 A = 1$ (section 65)

then $\sin^2 A = 1 - \cos^2 A$

and $\cos^2 A = 1 - \sin^2 A$

Substituting for $\cos^2 A$ in (4)

Substituting for $\sin^2 A$ $\cos 2A = 1 - 2 \sin^2 A \quad (5)$

$$\cos 2A = 2 \cos^2 A - 1 \quad (6)$$

No. 5 may be written in the form:

$$1 - \cos 2A = 2 \sin^2 A \quad (7)$$

and No. 6 as

$$1 + \cos 2A = 2 \cos^2 A \quad (8)$$

These alternative forms are very useful.

Again, if (7) be divided by (8)

$$\frac{1 - \cos 2A}{1 + \cos 2A} = \frac{\sin^2 A}{\cos^2 A}$$

or

$$\tan^2 A = \frac{1 - \cos 2A}{1 + \cos 2A} \quad (9)$$

If 2A be replaced by θ, formulae (4). (5) and (6) may be written in the forms

$$\cos \theta = \cos^2 \frac{\theta}{2} - \sin^2 \frac{\theta}{2} \quad (10)$$

$$\cos \theta = 1 - 2 \sin^2 \frac{\theta}{2} \tag{11}$$

$$\cos \theta = 2 \cos^2 \frac{\theta}{2} - 1 \tag{12}$$

**84** Similar formulae may be found for tangents.

Since $$\tan (A + B) = \frac{\tan A + \tan B}{1 - \tan A \tan B}$$

Let $B = A$

Then $$\tan 2A = \frac{2 \tan A}{1 - \tan^2 A} \tag{3}$$

or replacing 2A by $\theta$

$$\tan \theta = \frac{2 \tan \frac{\theta}{2}}{1 - \tan^2 \frac{\theta}{2}} \tag{14}$$

Formula (11) above may be written in the form:

$$\sin^2 \frac{\theta}{2} = \tfrac{1}{2} (1 - \cos \theta)$$

It is frequently used in navigation.

$(1 - \cos \theta)$ is called the **versed** sine of $\theta$
and $(1 - \sin \theta)$ is called the **coversed** sine of$\theta$.
$\frac{1}{2} (1 - \cos \theta)$ is called the **haversine**, i.e. half the versed sine.

**85** The preceding formulae are collected here for future reference.

(1) $\sin (A + B) = \sin A \cos B + \cos A \sin B$
(2) $\sin (A - B) = \sin A \cos B - \cos A \sin B$
(3) $\cos (A + B) = \cos A \cos B - \sin A \sin B$
(4) $\cos (A - B) = \cos A \cos B + \sin A \sin B$

(5) $\tan (A + B) = \dfrac{\tan A + \tan B}{1 - \tan A \tan B}$

(6) $\tan (A - B) = \dfrac{\tan A - \tan B}{1 + \tan A \tan B}$

(7) $\sin 2A \quad = 2 \sin A \cos A$

(8) $\cos 2A \quad = \cos^2 A - \sin^2 A$

$= 1 - 2 \sin^2 A$

$= 2 \cos^2 A - 1$

(9) $\tan 2A \quad = \dfrac{2 \tan A}{1 - \tan^2 A}$

These formulae should be carefully memorised. Variations of (7), (8), (9) in the form $\theta$ and $\frac{\theta}{2}$ should also be remembered.

*Exercise 8*

**1** If $\sin A = \frac{3}{5}$, find $\sin 2A$, $\cos 2A$ and $\tan 2A$.

**2** Find $\sin 2\theta$, $\cos 2\theta$. $\tan 2\theta$, when $\sin \theta = 0.25$.

**3** Given the values of $\sin 45°$ and $\cos 45°$ deduce the values of $\sin 90°$ and $\cos 90°$ by using the above formulae.

**4** If $\cos B = 0.66$, find $\sin 2B$ and $\cos 2B$.

**5** Find the values of (1) $2 \sin 36° \cos 36°$.
(2) $2 \cos^2 36° - 1$.

**6** If $\cos 2A = \frac{3}{5}$, find $\tan A$.
(*Hint* Use formulae from section 83.)

**7** Prove that
$$\sin \frac{\theta}{2} = \pm \sqrt{\frac{1 - \cos \theta}{2}}$$
$$\cos \frac{\theta}{2} = \pm \sqrt{\frac{1 + \cos \theta}{2}}$$

**8** If $\cos \theta = \frac{1}{2}$, find $\sin \frac{\theta}{2}$ and $\cos \frac{\theta}{2}$.

(*Hint* Use the rsults of the previous question.)

**9** If $1 - \cos 2\theta = 0.72$, find $\sin \theta$ and check by using the tables.

**10** Prove that $\cos^4 \theta - \sin^4 \theta = \cos 2\theta$.
(*Hint* Factorise the left-hand side.)

**11** Prove that $\left(\sin \frac{\theta}{2} + \cos \frac{\theta}{2}\right)^2 - 1 = \sin \theta$.

**12** Find the value of $\sqrt{\dfrac{1 - \cos 30°}{1 + \cos 30°}}$.

(*Hint* See formula from section 83.)

## 86 Product formulae

The formulae of section 80 give rise to another set of results involving the product of trigonometrical ratios.

We have seen that:

$$\sin(A+B) = \sin A \cos B + \cos A \sin B \quad (1)$$
$$\sin(A-B) = \sin A \cos B - \cos A \sin B \quad (2)$$
$$\cos(A+B) = \cos A \cos B - \sin A \sin B \quad (3)$$
$$\cos(A-B) = \cos A \cos B + \sin A \sin B \quad (4)$$

Adding (1) and (2)

$$\sin(A+B) + \sin(A-B) = 2\sin A \cos B$$

Subtracting

$$\sin(A+B) - \sin(A-B) = 2\cos A \sin B$$

Adding (3) and (4)

$$\cos(A+B) + \cos(A-B) = 2\cos A \cos B$$

Subtracting

$$\cos(A+B) - \cos(A-B) = -2\sin A \sin B$$

These can be written in the forms

$$2\sin A \cos B = \sin(A+B) + \sin(A-B) \quad (5)$$
$$2\cos A \sin B = \sin(A+B) - \sin(A-B) \quad (6)$$
$$2\cos A \cos B = \cos(A+B) + \cos(A-B) \quad (7)$$
$$2\sin A \sin B = \cos(A-B) - \cos(A+B) \quad (8)$$

*Note* The order on the right-hand side of (8) must be carefully noted.

## 87

Let $\quad A + B = P$

and $\quad A - B = Q$

Adding $\quad 2A = P + Q$

Subtracting $\quad 2B = P - Q$

$$\therefore \quad A = \frac{P+Q}{2}$$

$$B = \frac{P-Q}{2}$$

Substituting in (5), (6), (7) and (8)

$$\sin P + \sin Q = 2\sin\frac{P+Q}{2}\cos\frac{P-Q}{2} \quad (9)$$

$$\sin P - \sin Q = 2\cos\frac{P+Q}{2}\sin\frac{P-Q}{2} \quad (10)$$

$$\cos P + \cos Q = 2 \cos \frac{P + Q}{2} \cos \frac{P - Q}{2} \qquad (11)$$

$$\cos Q - \cos P = 2 \sin \frac{P + Q}{2} \sin \frac{P - Q}{2} \qquad (12)$$

The formulae (5), (6), (7), (8) enable us to change the product of two ratios into a sum.

Formulae (9), (10), (11), (12) enable us to change the sum of two ratios into a product.

Again note carefully the order in (12).

## 88 Worked examples

*Example 1*: Express as the sum of two trigonometrical ratios $\sin 5\theta \cos 3\theta$.

Using $2 \sin A \cos B = \sin (A + B) + \sin (A - B)$ on substitution

$$\begin{aligned} \sin 5\theta \cos 3\theta &= \tfrac{1}{2} \{\sin (5\theta + 3\theta) + \sin (5\theta - 3\theta)\} \\ &= \tfrac{1}{2} \{\sin 8\theta + \sin 2\theta\} \end{aligned}$$

*Example 2*: Change into a sum $\sin 70° \sin 20°$.

Using

$$2 \sin A \sin B = \cos (A - B) - \cos (A + B)$$

on substitution

$$\begin{aligned} \sin 70° \sin 20° &= \tfrac{1}{2} \{\cos (70° - 20°) - \cos (70° + 20°)\} \\ &= \tfrac{1}{2} \{\cos 50° - \cos 90°\} \\ &= \tfrac{1}{2} \cos 50° \quad \text{since } \cos 90° = 0 \end{aligned}$$

*Example 3*: Transform into a product $\sin 25° + \sin 18°$.

Using

$$\sin P + \sin Q = 2 \sin \frac{P + Q}{2} \cos \frac{P - Q}{2}$$

$$\sin 25° + \sin 18° = 2 \sin \frac{25° + 18°}{2} \cos \frac{25° - 18°}{2}$$

$$= 2 \sin 21.5° \cos 3.5°$$

*Example 4*: Change into a product $\cos 3\theta - \cos 7\theta$.

Using

$$\cos Q - \cos P = 2 \sin \frac{P + Q}{2} \sin \frac{P - Q}{2}$$

on substitution

$$\cos 3\theta - \cos 7\theta = 2 \sin \frac{3\theta + 7\theta}{2} \sin \frac{7\theta - 3\theta}{2}$$

$$= 2 \sin 5\theta \sin 2\theta$$

*Exercise 9*

Express as the **sum** or difference of two ratios:

**1** $\sin 3\theta \cos \theta$
**2** $\sin 35° \cos 45°$
**3** $\cos 50° \cos 30°$
**4** $\cos 5\theta \sin 3\theta$
**5** $\cos (C + 2D) \cos (2C + D)$
**6** cps $60° \sin 30°$
**7** $2 \sin 3A \sin A$
**8** $\cos (3C + 5D) \sin (3C - 5D)$

Express as the product of two ratios:

**9** $\sin 4A + \sin 2A$
**10** $\sin 5A - \sin A$
**11** $\cos 4\theta - \cos 2\theta$
**12** $\cos A - \cos 5A$
**13** $\cos 47° + \cos 35°$
**14** $\sin 49° - \sin 23°$

**15** $\dfrac{\sin 30° + \sin 60°}{\cos 30° - \cos 60°}$

**16** $\dfrac{\sin \alpha + \sin \beta}{\cos \alpha + \cos \beta}$

# 7

# Relations Between the Sides and Angles of a Triangle

**89** In section 61 we considered the relations which exist between the sides and angles of a right-angled triangle. In this chapter we proceed to deal similarly with any triangle.

In accordance with the usual practice, the angles of a triangle will be denoted by A, B, and C, and sides opposite to these by a, b, and c, respectively.

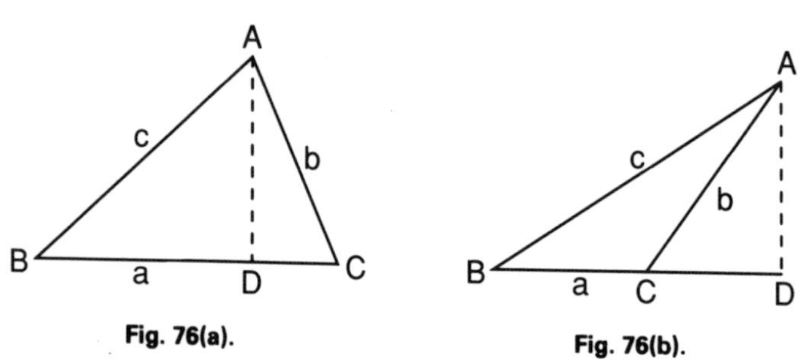


Fig. 76(a). Fig. 76(b).

## 90 The sine rule

In every triangle the sides are proportional to the sines of the opposite angles.

There are two cases to be considered:

(1) Acute-angled triangle (Fig. 76(a)).
(2) Obtuse-angled triangle (Fig. 76(b)).

In each figure draw AD perpendicular to BC, or to BC produced (Fig. 76(b)).

In $\triangle ABD$, $AD = c \sin B$ (1)

In $\triangle ACD$, $AD = b \sin C$ (2)

In Fig. 76(b), since ACB and ACD are supplementary angles

$$\sin ACD = \sin ACB = \sin C$$

Equating (1) and (2):

$$c \sin B = b \sin C$$

$$\therefore \quad \frac{b}{c} = \frac{\sin B}{\sin C}$$

Similarly $$\frac{a}{b} = \frac{\sin A}{\sin B}$$

and $$\frac{a}{c} = \frac{\sin A}{\sin C}$$

These results may be combined in the one formula

$$\frac{\sin A}{a} = \frac{\sin B}{b} = \frac{\sin C}{c}$$

These formulae are suitable for logarithmic calculations.

*Worked example* If in a triangle ABC, $A = 52.25°$, $B = 70.43°$ and $a = 9.8$, find b and c.

Using the sine rule:

$$\frac{b}{a} = \frac{\sin B}{\sin A}$$

$$\therefore \quad b = \frac{a \sin B}{\sin A}$$

$$= \frac{9.8 \times \sin 70.43°}{\sin 52.25°}$$

$$= 11.68$$

$$\therefore \quad b = 11.7 \text{ (approx.)}$$

On your calculator the sequence of key presses should be;
9.8 × 70.43 SIN ÷ 52.25 SIN =

Similarly c may be found by using $\frac{c}{a} = \frac{\sin C}{\sin A}$

*Exercise 10*

Solve the following problems connected with a triangle ABC.

**1** When A = 54°, B = 67°, a = 13.9 m, find b and c.
**2** When A = 38.25°, B = 29.63°, b = 16.2 m, find a and c.
**3** When A = 70°, C = 58.27°, b = 6 mm, find a and c.
**4** When A = 88°, B = 36°, a = 9.5 m, find b and c.
**5** When B = 75°, C = 42°, b = 25 cm, find a and c.

## 91 The cosine rule

As in the case of the sine rule, there are two cases to be considered. These are shown in Figs. 77(a) and 77(b).

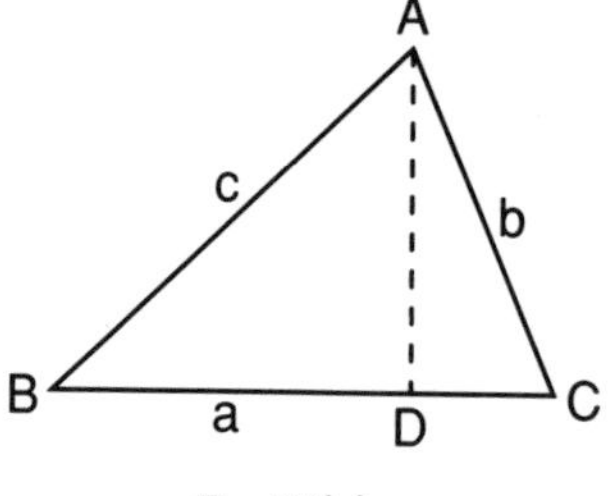


**Fig. 77(a).**

**Fig. 77(b).**

Let $\quad BD = x$

Then $\quad CD = a - x$ in Fig. 77(a)

and $\quad CD = x - a$ in Fig. 77(b)

In $\triangle ABD$, $\quad AD^2 = AB^2 - BD^2$

$$= c^2 - x^2 \qquad (1)$$

In $\triangle ACD$, $\quad AD^2 = AC^2 - CD^2$

$$= b^2 - (a - x)^2 \text{ in Fig. 77(a)} \qquad (2)$$

or
$$= b^2 - (x - a)^2 \text{ in Fig. 77(b)}$$

Also $\quad (a - x)^2 = (x - a)^2$

$\therefore$ equating (1) and (2)

$$b^2 - (a - x)^2 = c^2 - x^2$$

$$\therefore \quad b^2 - a^2 - 2ax - x^2 = c^2 - x^2$$

$$\therefore \quad 2ax = a^2 - c^2 - b^2$$

But $\quad x = c \cos B$

$$\therefore \quad 2ac \cos B = a^2 + c^2 - b^2$$

$$\therefore \quad \cos B = \frac{a^2 + c^2 - b^2}{2ac}$$

Similarly
$$\cos A = \frac{b^2 + c^2 - a^2}{2bc}$$

$$\cos C = \frac{a^2 + b^2 - c^2}{2ab}$$

The formulae may also be written in the forms:

$$c^2 = a^2 + b^2 - 2ab \cos C$$
$$a^2 = b^2 + c^2 - 2bc \cos A$$
$$b^2 = a^2 + c^2 - 2ac \cos B$$

These formulae enable us to find the angles of a triangle when all the sides are known. In the second form it enables us to find the third side when two sides and the enclosed angle are known.

*Worked example*

Find the angles of the triangle whose sides are

$$a = 8 \text{ m}, b = 9 \text{ m}, c = 12 \text{ m}.$$

Using
$$\cos C = \frac{a^2 + b^2 - c^2}{2ab}$$

$$= \frac{8^2 + 9^2 - 12^2}{2 \times 8 \times 9}$$

$$= \frac{64 + 81 - 144}{2 \times 8 \times 9}$$

$$= \frac{1}{144}$$

$$= 0.0069$$

whence
$$C = 89.6°$$

Again,
$$\cos A = \frac{b^2 + c^2 - a^2}{2bc}$$

$$= \frac{9^2 + 12^2 - 8^2}{2 \times 9 \times 12}$$

$$= \frac{81 + 144 - 64}{2 \times 9 \times 12}$$

$$= \frac{161}{216}$$

$$= 0.7454$$

whence
$$A = 41.8°$$

Similarly, using

$$\cos B = \frac{a^2 + c^2 - b^2}{2ac}$$

we get $$B = 48.6°$$

Check

$$\begin{aligned} &A + B + C \\ &= 41.8° + 48.6° + 89.6° \\ &= 180° \end{aligned}$$

*Exercise 11*

Find the angles of the triangles in which:

**1** a = 2 km, b = 3 km, c = 4 km.
**2** a = 54 mm, b = 71 mm, c = 83 mm.
**3** a = 24 m, b = 19 m, c = 26 m.
**4** a = 2.6 km, b = 2.85 km, c = 4.7 km.
**5** If a = 14 m, b = 8.5 m, c = 9 m, find the greatest angle of the triangle.
**6** When a = 64 mm, b = 57 mm, and c = 82 mm, find the smallest angle of the triangle.

## 92 The half-angle formulae

The cosine formula is tedious when the numbers involved are large: it is the basis, however, of a series of other formulae which are easier to manipulate.

## 93 To express the sines of half the angles in terms of the sides

As proved in section 91

$$\cos A = \frac{b^2 + c^2 - a^2}{2bc}$$

but $$\cos A = 1 - 2\sin^2\frac{A}{2} \qquad \text{(section 83)}$$

$$\therefore \quad 1 - 2\sin^2\frac{A}{2} = \frac{b^2 + c^2 - a^2}{2bc}$$

$$\therefore \quad 2\sin^2\frac{A}{2} = 1 - \frac{b^2 + c^2 - a^2}{2bc}$$

$$= \frac{2bc - (b^2 + c^2 - a^2)}{2bc}$$

$$= \frac{2bc - b^2 - c^2 + a^2}{2bc}$$

$$= \frac{a^2 - (b^2 - 2bc + c^2)}{2bc}$$

$$= \frac{a^2 - (b - c)^2}{2bc}$$

Factorising the numerator

$$2 \sin^2 \frac{A}{2} = \frac{(a + b - c)(a - b + c)}{2bc} \qquad \text{(A)}$$

The 's' notation. To simplify this further we use the 's' notation, as follows:

Let $2s = a + b + c$, i.e. the perimeter of the triangle.

Then $\quad 2s - 2a = a + b + c = 2a$
$\qquad\qquad = b + c - a$

Again $\quad 2s - 2b = a + b + c - 2b$
$\qquad\qquad = a - b + c$

Similarly $\quad 2s - 2c = a + b - c$

These may be written

$$2s = a + b + c \qquad (1)$$
$$2(s - a) = b + c - a \qquad (2)$$
$$2(s - b) = a - b + c \qquad (3)$$
$$2(s - c) = a + b - c \qquad (4)$$

From (A) above

$$2 \sin^2 \frac{A}{2} = \frac{(a + b - c)(a - b + c)}{2bc}$$

Replacing the factors of the numerator by their equivalents in formulae (3) and (4)

we have $$2 \sin^2 \frac{A}{2} = \frac{2(s - c) \times 2(s - b)}{2bc}$$

Cancelling the 2's.

$$\sin^2 \frac{A}{2} = \frac{(s - c)(s - b)}{bc}$$

or $$\sin\frac{A}{2} = \sqrt{\frac{(s-b)(s-c)}{bc}}$$

Similarly, $$\sin\frac{B}{2} = \sqrt{\frac{(s-a)(s-c)}{ac}}$$

$$\sin\frac{C}{2} = \sqrt{\frac{(s-a)(s-b)}{ab}}$$

## 94 To express the cosines of half the angles of a triangle in terms of the sides

Since $$\cos A = \frac{b^2 + c^2 - a^2}{2bc}$$

$$1 + \cos A = 1 + \frac{b^2 + c^2 - a^2}{2bc}$$

but $$1 + \cos A = 2\cos^2\frac{A}{2} \qquad \text{(chapter 6, section 83)}$$

$$\therefore \quad 2\cos^2\frac{A}{2} = 1 + \frac{b^2 + c^2 - a^2}{2bc}$$

$$= \frac{(b^2 + 2bc + c^2) - a^2}{2bc}$$

$$= \frac{(b + c)^2 - a^2}{2bc}$$

$$= \frac{(b + c - a)(b + c + a)}{2bc}$$

(on factorising the numerator)

but $$b + c - a = 2(s - a)$$
and $$a + b + c = 2s$$

Substituting

$$2\cos^2\frac{A}{2} = \frac{2(s-a) \times 2s}{2bc}$$

and $$\cos^2\frac{A}{2} = \frac{s(s-a)}{bc}$$

$$\therefore \quad \cos\frac{A}{2} = \sqrt{\frac{s(s-a)}{bc}}$$

Similarly $$\cos\frac{B}{2} = \sqrt{\frac{s(s-b)}{ac}}$$

$$\cos\frac{C}{2} = \sqrt{\frac{s(s-c)}{ab}}$$

## 95 To express the tangents of half the angles of a triangle in terms of the sides

Since $$\tan\frac{A}{2} = \frac{\sin\frac{A}{2}}{\cos\frac{A}{2}}$$

we can substitute for $\sin\frac{A}{2}$ and $\cos\frac{A}{2}$ the expressions found above.

Then $$\tan\frac{A}{2} = \frac{\sqrt{\frac{(s-b)(s-c)}{bc}}}{\sqrt{\frac{s(s-a)}{bc}}}$$

Simplifying and cancelling

$$\tan\frac{A}{2} = \sqrt{\frac{(s-b)(s-c)}{s(s-a)}}$$

Similarly $$\tan\frac{B}{2} = \sqrt{\frac{(s-a)(s-c)}{s(s-b)}}$$

and $$\tan\frac{C}{2} = \sqrt{\frac{(s-a)(s-b)}{s(s-c)}}$$

## 96 To express the sine of an angle of a triangle in terms of the sides

Since

$$\sin A = 2\sin\frac{A}{2}\cos\frac{A}{2}$$

substituting for $\sin\frac{A}{2}$ and $\cos\frac{A}{2}$ the values found above

$$\sin A = 2\sqrt{\frac{(s-b)(s-c)}{bc}} \times \sqrt{\frac{s(s-a)}{bc}}$$

$$\therefore \quad \sin A = \frac{2}{bc}\sqrt{s(s-a)(s-b)(s-c)},$$ on simplifying.

Similarly

$$\sin B = \frac{2}{ac}\sqrt{s(s-a)(s-b)(s-c)}$$

and

$$\sin C = \frac{2}{ab}\sqrt{s(s-a)(s-b)(s-c)}$$

## 97 Worked example

The working involved in the use of all these formulae is very similar. We will give one example only: there are more in the next chapter.

*The sides of a triangle are a = 264, b = 435, c = 473. Find the greatest angle.*

The greatest angle is opposite to the greatest side and is therefore C.

Begin by calculating values of the 's' factors.

$$\begin{aligned} a &= 264 \\ b &= 435 \\ c &= 473 \\ \hline \therefore \quad 2s &= 1172 \\ \hline s &= 586 \\ s - a &= 322 \\ s - b &= 151 \\ s - c &= 113 \\ \hline 2s &= 1172 \end{aligned}$$

and (for s and the s-factors); Check (for the final 2s).

*Note* $s + (s-a) + (s-b) + (s-c) = 4s - (a+b+c) = 2s$

Any of the half angle formulae may be used, but the tangent formulae involves only the 's' factors.

Using

$$\tan\frac{C}{2} = \sqrt{\frac{(s-c)(s-b)}{s(s-c)}}$$

$$\tan\frac{C}{2} = \sqrt{\frac{322 \times 151}{586 \times 113}}$$

$= 0.8569$
$= \tan (40.59°)$

$$\therefore \quad \frac{C}{2} = 40.59°$$

and $\quad C = 81.19°$

On your calculator the sequence of key presses should be:
322 × 151 ÷ 586 ÷ 113 = √INV TAN × 2 =

*Exercise 12*

**1** Using the formula for $\tan \frac{A}{2}$, find the largest angle in the triangle whose sides are 113 mm, 141 mm, 214 mm.

**2** Using the formula for $\sin \frac{A}{2}$, find the smallest angle in the triangle whose sides are 483 mm, 316 mm, and 624 mm.

**3** Using the formula for $\cos \frac{B}{2}$ find B when $a = 115$ m, $b = 221$ m, $c = 286$ m.

**4** Using the half-angle formulae find the angles of the triangle when $a = 160$, $b = 220$, $c = 340$.

**5** Using the half-angle formulae find the angles of the triangle whose sides are 73.5, 65.5 and 75.

**6** Using the formula for the sine in section 96 find the smallest angle of the triangle whose sides are 172 km, 208 km, and 274 km.

## 98 To prove that in any triangle

$$\tan \frac{B - C}{2} = \frac{b - c}{b + c} \cot \frac{A}{2}$$

From section 90 $\quad \frac{\sin B}{b} = \frac{\sin C}{c}$

Let each of these ratios equal k.

Then $\quad \sin B = bk \qquad (1)$

and $\quad \sin C = ck \qquad (2)$

Adding (1) and (2)

$$\sin B + \sin C = k(b + c) \qquad (3)$$

Subtracting (2) from (1)

$$\sin B - \sin C = k(b - c) \qquad (4)$$

Dividing (4) by (3)

$$\frac{\sin B - \sin C}{\sin B + \sin C} = \frac{b - c}{b + c}$$

or

$$\frac{b - c}{b + c} = \frac{\sin B - \sin C}{\sin B + \sin C}$$

Applying to the numerator and denominator of the right-hand side the formulae, 9 and 10 of section 87.

We get

$$\frac{b - c}{b + c} = \frac{2 \cos \frac{B + C}{2} \cdot \sin \frac{B - C}{2}}{2 \sin \frac{B + C}{2} \cdot \cos \frac{B - C}{2}}$$

$$= \frac{\sin \frac{B - C}{2}}{\cos \frac{B - C}{2}} \div \frac{\sin \frac{B + C}{2}}{\cos \frac{B + C}{2}}$$

$$= \frac{\tan \frac{B - C}{2}}{\tan \frac{B + C}{2}}$$

Since

$$(B + C) = 180° - A$$

$$\therefore \quad \frac{B + C}{2} = 90° - \frac{A}{2}$$

$$\therefore \quad \frac{b - c}{b + c} = \frac{\tan \frac{B - C}{2}}{\tan \left(90° - \frac{A}{2}\right)}$$

$$= \frac{\tan \frac{B - C}{2}}{\cot \frac{A}{2}} \qquad \text{(see section 53)}$$

$$\therefore \quad \frac{\tan \frac{B - C}{2}}{\cot \frac{A}{2}} = \frac{b - c}{b + c}$$

or

$$\tan \frac{B - C}{2} = \frac{b - c}{b + c} \cot \frac{A}{2}$$

Similarly

$$\tan\frac{A - C}{2} = \frac{a - c}{a + c}\cot\frac{B}{2}$$

$$\tan\frac{A - B}{2} = \frac{a - b}{a + b}\cot\frac{C}{2}$$

On the right-hand side we have quantities which are known when we are given *two sides of a triangle and the contained angle.*

Consequently we can find $\frac{B - C}{2}$ and so B − C.

Since A is known we can find B + C for B + C = 180 − A

| | | |
|---|---|---|
| Let | $B + C = \alpha$ | |
| Let | $B - C = \beta$ | (note $\alpha$ and $\beta$ are now known) |
| Adding | $2B = \alpha + \beta$ | |
| Subtracting | $2C = \alpha - \beta$ | |

$$\therefore \quad B = \frac{\alpha + \beta}{2} \text{ and } C = \frac{\alpha - \beta}{2}$$

Hence we know all the angles of the triangle.

## *Worked example*

*In a triangle A = 75.2°, b = 43, c = 35. Find B and C.*

Using $$\tan\frac{B - C}{2} = \frac{b - c}{b + c}\cot\frac{A}{2}$$

and substituting

$$\tan\frac{B - C}{2} = \frac{43 - 35}{43 + 35}\cot 37.6°$$

$$= \frac{8}{78}\cot 37.6°$$

$$= \frac{8}{78} \div \tan 37.6°$$

$$= 0.1332$$

$$= \tan(7.59°)$$

On your calculator the sequence of key presses should be:
(43 − 35) ÷ (43 + 35) ÷ 37.6 TAN = INV TAN

whence $$\frac{B - C}{2} = 7.59°$$

and $$B - C = 15.17°$$

Also $B + C = 180° - 75.2°$
$= 104.8°$

(1) Adding $2B = 119.97°$
and $B = 59.98°$
$\therefore B = 60°$ (approx.)

(2) Subtracting $2C = 89.63°$
and $C = 44.86°$

## 99 To prove that in any triangle

$$a = b \cos C + c \cos B$$

As in section 90 there are two cases.

In Fig. 78(a) $BC = BD + DC$
But $BD = c \cos B$
and $DC = b \cos C$

$$\therefore a = BD + DC$$
$$= c \cos B + b \cos C$$

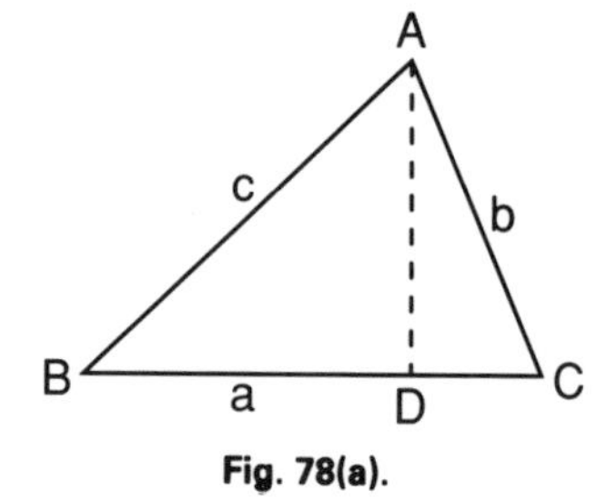


**Fig. 78(a).**

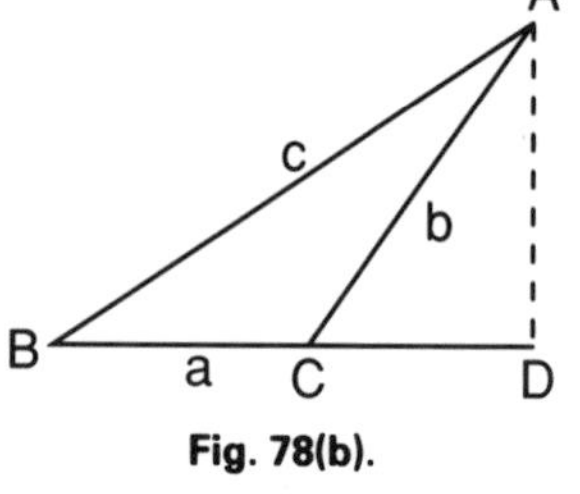


**Fig. 78(b).**

In Fig.78(b) $BC = BD - DC$

$$\therefore a = c \cos B - b \cos ACD$$
$$= c \cos B - b \cos (180° - C)$$
$$= c \cos B + b \cos C$$

since $\cos (180° - B) = -\cos B$ (see section 70)
$\therefore$ in each case

$$a = b \cos C + c \cos B$$

Similarly $b = a \cos C + c \cos A$
and $c = a \cos B + b \cos A$

Referring to section 63 we see that BD is the projection of AB on BC, and DC is the projection of AC on BC; in the second case BC is produced and the projection must be regarded as negative. Hence we may state the Theorem thus:

*Any side of a triangle is equal to the projection on it of the other two sides.*

*Exercise 13*

Use the formula provided in section 98 to find the remaining angles of the following triangles.

**1** a = 171, c = 288, B = 108°
**2** a = 786, b = 854, C = 37.42°
**3** c = 175, b = 602, A = 63.67°
**4** a = 185, b = 111, C = 60°
**5** a = 431, b = 387, C = 29.23°
**6** a = 759, c = 567, B = 72.23°

# 8

# The Solution of Triangles

**100** The formulae which have been proved in the previous chapter are those which are used for the purpose of **solving a triangle**. By this is meant that, given certain of the sides and angles of a triangle, we proceed to find the others. The parts given must be such as to make it possible to determine the triangle uniquely. If, for example, all the angles are given, there is no one triangle which has these angles, but an infinite number of such triangles, with different lengths of corresponding sides. Such triangles are **similar**, but not **congruent** (see section 15).

The conditions under which the solution of a triangle is possible must be the same as those which determine when triangles are congruent. If you need to revise these conditions turn back to section 13.

It should be understood, of course, that we are not dealing now with right-angled triangles, which have already been considered (see chapter 3, section 62).

**101** From the Theorems enumerated in section 13, it is clear that a triangle can be **solved** when the following parts are given:

**Case I** **Three sides**
**Case II** **Two sides and an included angle**
**Case III** **Two angles and a side**
**Case IV** **Two sides and an angle opposite to one of them**

This last case is the **ambiguous case** (see section 13) and under certain conditions, which will be dealt with later, there may be two solutions.

In the previous chapter, after proving the various formulae, examples were considered which were, in effect, concerned with the solution of a triangle, but we must now proceed to a systematic consideration of the whole problem.

## 102 Case I To solve a triangle when three sides are known

The problem is that of finding at least two of the angles, because since the sum of the angles of a triangle is 180°, when two are known the third can be found by subtraction. It is better, however, to calculate all three angles separately and check the result by seeing if their sum is 180°.

### Formulae employed

***(1) The cosine rule***
The formula

$$\cos A = \frac{b^2 + c^2 - a^2}{2bc}$$

will give A, and B and C can be similarly determined.

***(2) The half angle formulae***
The best of these, as previously pointed out, is the $\tan \frac{A}{2}$ formula, viz.

$$\tan \frac{A}{2} = \sqrt{\frac{(s-b)(s-c)}{s(s-a)}}$$

However, the formulae for $\sin \frac{A}{2}$ and $\cos \frac{A}{2}$ may be used.

***(3) The sine formula***

$$\sin A = \frac{2}{bc}\sqrt{s(s-a)(s-b)(s-c)}$$

This is longer than the half-angle formulae.

### *Worked example*

Solve the triangle in which a = 269.8, b = 235.9, c = 264.7.

Data

$$\begin{aligned} a &= 269.8 \\ b &= 235.9 \\ c &= 264.7 \\ \hline 2s &= 770.4 \\ \hline \therefore \quad s &= 385.2 \\ s - a &= 115.4 \\ s - b &= 149.3 \\ s - c &= 120.5 \\ \hline \end{aligned}$$

Check $\quad 2s = 770.4$

To find A

Formula to be used $\tan \frac{A}{2} = \sqrt{\frac{(s-b)(s-c)}{s(s-a)}}$

$$= \sqrt{\frac{149.3 \times 120.5}{385.2 \times 115.4}}$$
$$= 0.6362$$
$$= \tan 32.46°$$

$$\therefore \quad \frac{A}{2} = 32.46°$$

and $\quad A = 64.93°$

On your calculator the sequence of key presses should be:
149.3 × 120.5 ÷ 385.2 ÷ 115.4 = √ INV TAN × 2 =

To find B

Formula to be used $\tan \frac{B}{2} = \sqrt{\frac{(s-a)(s-c)}{s(s-b)}}$

$$= \sqrt{\frac{115.4 \times 120.5}{385.2 \times 149.3}}$$
$$= 0.4917$$
$$= \tan 26.18°$$

$$\therefore \quad \frac{B}{2} = 26.18°$$

and $\quad B = 52.37°$

On your calculator the sequence of key presses should be:
115.4 × 120.5 ÷ 385.2 ÷ 149.3 = √ INV TAN × 2 =

To find C

Formula to be used $\tan\frac{C}{2} = \sqrt{\frac{(s-a)(s-b)}{s(s-c)}}$

$$= \sqrt{\frac{115.4 \times 149.3}{385.2 \times 120.5}}$$
$$= 0.6093$$
$$= \tan 31.35°$$

$$\therefore \quad \frac{C}{2} = 31.35°$$

and $\qquad C = 62.70°$

On your calculator the sequence of key presses should be:
115.4 × 149.3 ÷ 385.2 ÷ 120.5 = √ INV TAN × 2 =

Check

$$A = 64.93°$$
$$B = 52.37°$$
$$C = 62.70°$$

$$A + B + C = 180°$$

*Exercise 14*

Solve the following triangles:

**1** a = 252, b = 342, c = 486.
**2** a = 20, b = 11, c = 12.
**3** a = 206.5, b = 177, c = 295.
**4** a = 402.5, b = 773.5, c = 1001.
**5** a = 95.2, b = 162.4, c = 117.6.

## 103 Case II Given two sides and the contained angle

(1) The **cosine rule** may be used. If, for example, the given sides are b and c and the angle A, then

$$a^2 = b^2 + c^2 - 2bc \cos A$$

will give a.

Hence, since all sides are now known we can proceed as in Case I.

(2) Use the formula

$$\tan\frac{B-C}{2} = \frac{b-c}{b+c}\cot\frac{A}{2}$$

Solve the triangle when

$$b = 294,\ c = 406,\ A = 35.4°$$

Formula used:

$$\tan\frac{C - B}{2} = \frac{c - b}{c + b}\cot\frac{A}{2}$$

Data

$$\begin{aligned} b &= 294 \\ c &= 406 \\ c + b &= 700 \\ {}^{*}\ c - b &= 112 \\ A &= 35.4° \end{aligned}$$

$$\frac{A}{2} = 17.7°$$

$$\begin{aligned} C + B &= 144.6° \\ &= \frac{112}{700} \times \frac{1}{\tan 17.7°} \\ &= 0.5013 \\ &= \tan 26.63° \end{aligned}$$

$$\therefore \quad \frac{C - B}{2} = 26.63°$$

$$C - B = 53.25°$$

Also

$$\begin{aligned} C + B &= 144.6° \\ 2C &= 197.85° \\ C &= 98.92° \end{aligned}$$

Also

$$\begin{aligned} 2B &= 91.35° \\ B &= 45.68° \end{aligned}$$

To find a

Formula used:

$$\frac{a}{\sin A} = \frac{b}{\sin B}$$

$$a = \frac{b \sin A}{\sin B}$$

$$\begin{aligned} a &= 294 \times \sin 35.4° \div \sin 45.68° \\ &= 238.04 \end{aligned}$$

$$\therefore \quad a = 238 \text{ (approx.)}$$

* This form is used since $c > b$, and therefore $C > B$.

The solution is:

$$B = 45.68^\circ$$
$$C = 98.92^\circ$$
$$a = 238$$

*Exercise 15*

Solve the following triangles:

**1** b = 189, c = 117.7, A = 60.6°.
**2** a = 94, b = 159.4, C = 80.97°.
**3** a = 39.6, c = 71.1, B = 65.17°.
**4** a = 266, b = 175, C = 78°.
**5** a = 230.1, c = 269.5, B = 30.47°.

## 104 Case III Given two angles and a side

If two angles are known the third is also known, since the sum of all three angles is 180°. This case may therefore be stated as *given the angles and one side*.

The best formula to use is the **sine rule**.

### *Worked example*

Solve the triangle in which B = 71.32°, C = 67.45° and b = 79.06.

Required to find, A, a and c.

Now $$A = 180^\circ - 71.32^\circ + 67.45^\circ$$
$$= 41.23^\circ$$

To find c

Formula used $\frac{c}{b} = \frac{\sin C}{\sin B}$

whence $$c = \frac{b \sin C}{\sin B}$$

$$= 79.06 \times \sin 67.45^\circ \div \sin 71.32^\circ$$
$$= 77.08$$
$$c = 77.08$$

To find a

Using
$$\frac{a}{b} = \frac{\sin A}{\sin B}$$

$$a = \frac{b \sin A}{\sin B}$$

$$= 79.06 \times \sin 41.23° \div \sin 71.32°$$
$$= 55.00$$
$$\therefore \quad a = 55.00$$

$\therefore$ The solution is

$$A = 41.23°$$
$$a = 55.00$$
$$c = 77.08$$

*Exercise 16*

Solve the triangles:

**1** $a = 141.4$, $A = 74.3°$, $C = 24.23°$
**2** $b = 208.5$, $A = 95.68$, $B = 41.63°$
**3** $A = 29.93°$, $C = 108°$, $a = 112.8$
**4** $B = 32.68°$, $C = 49.63°$, $c = 117.6$
**5** $b = 11.74$, $A = 27.75°$, $B = 41.37°$

## 105 Case IV Given two sides and an angle opposite to one of them

This is the **ambiguous** case and you should revise chapter 1, section 13, before proceeding further.

As we have seen if the two sides and an angle opposite one of them are given, then the triangle is not always **uniquely** determined as in the previous cases, but there may be two solutions.

We will now consider from a trigonometrical point of view how this ambiguity may arise.

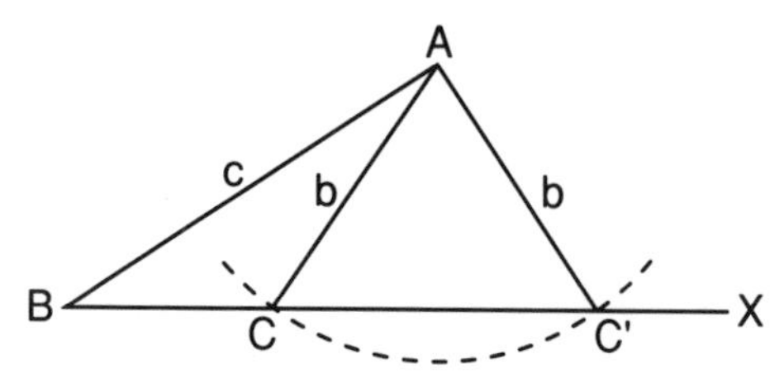


**Fig. 79.**

In the $\triangle ABC$ (Fig. 79), let c, b and B be known.

As previously shown in section 13 the side b may be drawn in two positions AC and AC′.

Both the triangles ABC and ABC′ satisfy the given conditions. Consequently there are:

(1) Two values for a, viz. BC and BC′.
(2) Two values for $\angle C$, viz. ACB or AC′B.
(3) Two values for $\angle A$, viz. BAC or BAC′.

Now the $\triangle ACC'$ is isosceles, since $AC = AC'$

$$\therefore \quad \angle ACC' = AC'C.$$

But ACC′ is the supplement of ACB.

$\therefore$ also AC′C is the supplement of ACB.

$\therefore$ the two *possible* values of $\angle C$, viz. ACB and AC′B are *supplementary*.

*Solution*

Since c, b, B are known, C can be found by the sine rule.

i.e. we use
$$\frac{\sin C}{c} = \frac{\sin B}{b}$$

whence
$$\sin C = \frac{c \sin B}{b}$$

Let us suppose that $c = 8.7$, $b = 7.6$, $B = 25$

Then
$$\sin C = \frac{8.7 \sin 25°}{7.6}$$
$$= 0.4838$$

We have seen in section 73 that when the value of a sine is given, there are two angles less than 180° which have that sine, and the angles are supplementary. The acute angle whose sine is 0.4638 is 28.93°.

Consequently there are two values for C, viz.

$$28.93° \text{ and } 157.07°$$

Let us examine the question further by considering the consequences of variations relative to c in the value of b, the side opposite to the given angle B.

As before draw BA making the given angle B meet BX, of indefinite length. Then with centre A and radius = b draw an arc of a circle.

(1) If this arc *touches* BX in C, we have the minimum length of b to make a triangle at all (Fig. 80(a)). The triangle is then right-angled, there is *no ambiguity* and

$$b = c \sin B$$

(2) If b is $> c \sin B$ but $< c$ then BX is cut in two points C and C′ (Fig. 80(b)).

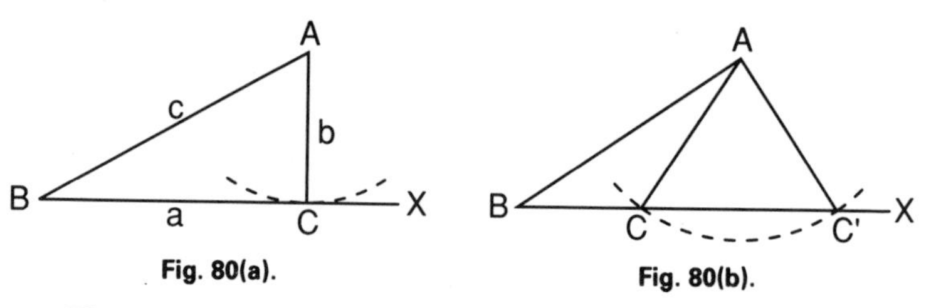


Fig. 80(a).

Fig. 80(b).

There are two $\triangle$s ABC, ABC′ and the case is *ambiguous*.

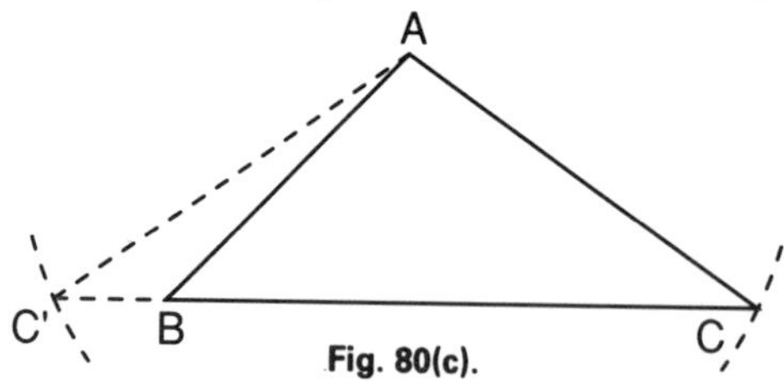


Fig. 80(c).

(3) If $b > c$, BX is cut at two points C and C′ (Fig. 80(c)), but one of these C′ lies on BX produced in the other direction and in the $\triangle$ so formed, there is no angle B, but only its supplement. There is one solution and no ambiguity.

$\therefore$ There are two solutions only when b, the side opposite to the given angle B, is less than c, the side adjacent, and greater than c sin B.

Ambiguity can therefore be ascertained by inspection.

*Exercise 17*

In the following cases ascertain if there is more than one solution. Then solve the triangles:

**1** $b = 30.4$, $c = 34.8$, $B = 25°$
**2** $b = 70.25$, $c = 85.3$, $B = 40°$
**3** $a = 96$, $c = 100$, $C = 66°$
**4** $a = 91$, $c = 78$, $C = 29.45°$

## 106 Area of a triangle

From many practical points of view, e.g. surveying, the calculation of the area of a triangle is an essential part of solving the triangle. This can be done more readily when the sides and angles are known. This will be apparent in the following formulae.

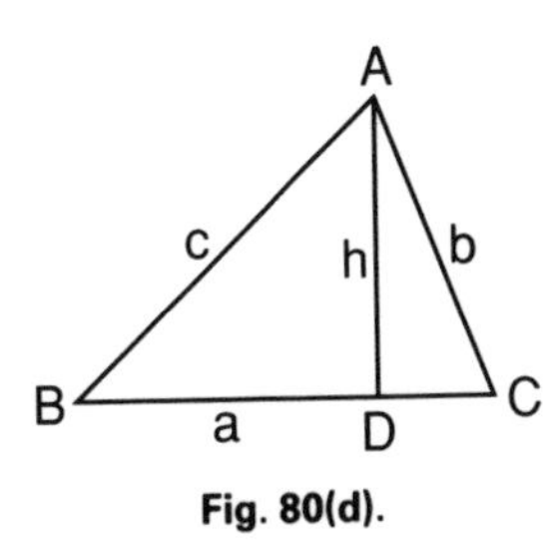


Fig. 80(d).

(1) *The base and altitude formula*

The student is probably acquainted with this formula which is easily obtained from elementary geometry.

Considering the triangle ABC in Fig. 80(d).

From A, a vertex of the triangle, draw AD perpendicular to the opposite side.

Let AD = h and let $\triangle$ = the area of the triangle.

Then
$$\triangle = \tfrac{1}{2}\,BC \times AD$$
$$= \tfrac{1}{2}\,ah$$

If perpendiculars are drawn from the other vertices B and C, similar formulae may be obtained.

It will be noticed that h is not calculated directly in any of the formulae for the solution of a triangle. It is generally more convenient to express it in terms of the sides and angle. Accordingly we modify this formula in (2).

(2) *The sine formula*

Referring to Fig. 80(d):

$$\frac{AD}{AC} = \sin C$$

$$\therefore \quad h = b \sin C$$

Substituting for h in formula above,

$$\triangle = \tfrac{1}{2}\,ab \sin C$$

Similarly using other sides as bases

$$\triangle = \tfrac{1}{2}\,bc \sin A$$
$$= \tfrac{1}{2}\,ac \sin B$$

This is a useful formula and adapted to logarithmic calculation.

It may be expressed as follows:

*The area of a triangle is equal to half the product of two sides and the sine of the angle contained by them.*

(3) *Area in terms of the sides*

We have seen in section 96, chapter 7, that

$$\sin A = \frac{2}{bc}\sqrt{s(s-a)(s-b)(s-c)}$$

Substituting this for sin A in the formula

$$\Delta = \tfrac{1}{2}bc \sin A$$

$$\Delta = \tfrac{1}{2}bc \times \frac{2}{bc}\sqrt{s(s-a)(s-b)(s-c)}$$

$$\therefore \quad \Delta = \sqrt{s(s-a)(s-b)(s-c)}$$

## Worked examples

(1) Find the area of the triangle solved in section 103, viz. $b = 294$, $c = 406$, $A = 35.4°$.

Using the formula:

$$\Delta = \tfrac{1}{2}bc \sin A$$
$$\Delta = \tfrac{1}{2} \times 294 \times 406 \times \sin 35.4°$$
$$\therefore \quad \Delta = 34570 \text{ sq. units}$$

(2) Find the area of the triangle solved in section 102, viz. $a = 269.8$, $b = 235.9$, $c = 264.7$.

Using the formula and taking values of s, s − a, etc., as in section 102:

$$\Delta = \sqrt{s(s-a)(s-b)(s-c)}$$
$$= \sqrt{385.2 \times 115.4 \times 149.3 \times 120.5}$$
$$= 28279.35$$
$$\therefore \quad \Delta = 28280 \text{ sq. units}$$

*Exercise 18*

**1** Find the area of the triangle when $a = 6.2$ m, $b = 7.8$ m, $C = 52°$.

**2** Find the area of the triangle ABC when AB = 14 km, BC = 11 km and $\angle ABC = 70°$.

**3** If the area of a triangle is 100 $m^2$ and two of its sides are 21 m and 15 m, find the angle between these sides.

**4** Find the area of the triangle when a = 98.2 cm, c = 73.5 cm and B = 135.33°.

**5** Find the area of the triangle whose sides are 28.7 cm, 35.4 cm and 51.8 cm.

**6** The sides of a triangle are 10 mm, 13 mm and 17 mm. Find its area.

**7** Find the area of the triangle whose sides are 23.22, 31.18 and 40.04 mm.

**8** Find the area of the triangle whose sides are 325 m, 256 m and 189 m.

**9** A triangle whose sides are 135 μm, 324 μm and 351 μm is made of material whose density is 2.3 kg $\mu m^{-2}$. Find the mass of the triangle in Mg.

**10** Find the area of a quadrilateral ABCD, in which AB = 14.7 cm, BC = 9.8 cm, CD = 21.7 cm, AD = 18.9 cm and ∠ABC = 137°.

**11** ABC is a triangle with sides BC = 36 cm, CA = 25 cm, AB = 29 cm. A point O lies inside the triangle and is distant 5 cm from BC and 10 cm from CA. Find its distance from AB.

## *Exercise 19*
## Miscellaneous Examples

**1** The least side of a triangle is 3.6 km long. Two of the angles are 37.25° and 48.4°. Find the greatest side.

**2** The sides of a triangle are 123 m, 79 m and 97 m. Find its angles as accurately as you can.

**3** Given b = 532.4, c = 647.1, A = 75.23°, find B, C and a.

**4** In a triangle ABC find the angle ACB when AB = 92 mm, BC = 50 mm and CA = 110 mm.

**5** The length of the side BC of a triangle ABC is 14.5 m, ∠ABC = 71°, ∠BAC = 57°. Calculate the lengths of the sides AC and AB.

**6** In a quadrilateral ABCD, AB = 3 m, BC = 4 m, CD = 7.4 m, DA = 4.4 m and the ∠ABC is 90°. Determine the angle ADC.

**7** When a = 25, b = 30, A = 50° determine how many such triangles exist and complete their solution.

**8** The length of the shortest side of a triangle is 162 m. If two angles are 37.25° and 48.4° find the greatest side.

**9** In a quadrilateral ABCD, AB = 4.3 m, BC = 3.4 m, CD = 3.8 m, $\angle ABC = 95°$, $\angle BCD = 115°$. Find the lengths of the diagonals.

**10** From a point O on a straight line OX, OP and OQ of lengths 5 mm and 7 mm are drawn on the same side of OX so that $\angle XOP = 32°$ and $\angle XOQ = 55°$. Find the length of PQ.

**11** Two hooks P and Q on a horizontal beam are 30 cm apart. From P and Q strings PR and QR, 18 cm and 16 cm long respectively, support a weight at R. Find the distance of R from the beam and the angles which PR and QR make with the beam.

**12** Construct a triangle ABC whose base is 5 cm long, the angle BAC = 55° and the angle ABC = 48°. Calculate the lengths of the sides AC and BC and the area of the triangle.

**13** Two ships leave port at the same time. The first steams S.E. at 18 km $h^{-1}$, and the second 25° W. of S. at 15 km $h^{-1}$. Calculate the time that will have elapsed when they are 86 km apart.

**14** AB is a base line of length 3 km, and C, D are points such that $\angle BAC = 32.25°$, $\angle ABC = 119.08°$, $\angle DBC = 60.17°$, $\angle BCD = 78.75°$, A and D being on the same side of BC. Prove that the length of CD is 4405 m approximately.

**15** ABCD is a quadrilateral. If AB = 0.38 m, BC = 0.69 m, AD = 0.42 m, $\angle ABC = 109°$, $\angle BAD = 123°$, find the area of the quadrilateral.

**16** A weight was hung from a horizontal beam by two chains 8 m and 9 m long respectively, the ends of the chains being fastened to the same point of the weight, their other ends being fastened to the beam at points 10 m apart. Determine the angles which the chains make with the beam.

# 9

# Practical Problems Involving the Solution of Triangles

**107** It is not possible within the limits of this book to deal with the many practical applications of trigonometry. For adequate treatment of these the student must consult the technical works specially written for those professions in which the subject is necessary. All that is attempted in this chapter is the consideration of a few types of problems which embody those principles which are common to most of the technical applications. Exercises are provided which will provide a training in the use of the rules and formulae which have been studied in previous chapters. In other words, you must learn to use your tools efficiently and accurately.

## 108 Determination of the height of a distant object

This problem has occupied the attention of mankind throughout the ages. Three simple forms of the problem may be considered here.

(a) When the point vertically beneath the top of the object is accessible

In Fig. 81 AB represents a lofty object whose height is required, and B is the foot of it, on the same horizontal level as O. This being accessible, a horizontal distance represented by OB can be measured. By the aid of a theodolite the angle of elevation of AB, viz. $\angle$AOB, can be found.

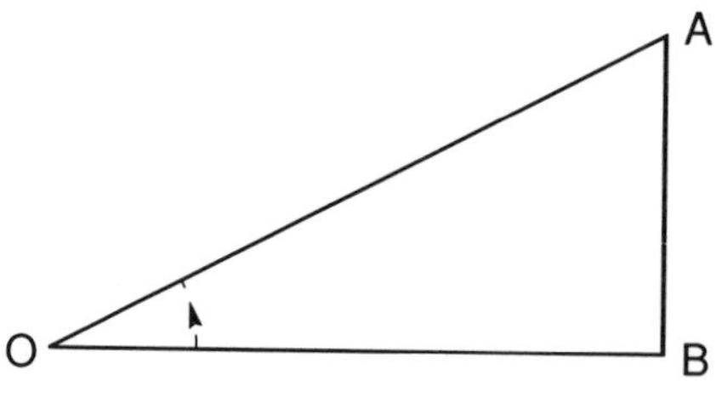


**Fig. 81.**

Then $$AB = PB \tan AOB$$

The case of the pyramid considered in chapter 3, section 40, is an example of this. It was assumed that distance from the point vertically below the top of the pyramid could be found.

(b) When the point on the ground vertically beneath the top of the object is not accessible

In Fig. 82 AB represents the height to be determined and B is not accessible. To determine AB we can proceed as follows:

From a suitable point Q, $\angle AQB$ is measured by means of a theodolite.

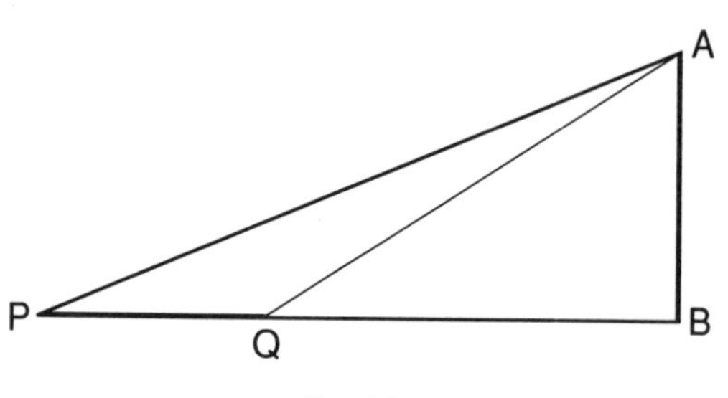


**Fig. 82.**

Then a distance PQ is measured so that P and Q are on the same horizontal plane as B and the $\triangle APQ$ and AB are in the same vertical plane.

Then $\angle APQ$ is measured.

$\therefore$ in $\triangle APQ$.

PQ is known.
$\angle APQ$ is known.
$\angle AQP$ is known, being the supplement of $\angle AQB$.

The $\triangle APQ$ can therefore be solved as in Case 3, section 104.
When AP is known.

Then $\quad AB = AP \sin APB$
As a check $\quad AB = AQ \sin AQB$

(c) By measuring a horizontal distance in any direction

It is not always easy to obtain a distance PQ as in the previous example, so that $\triangle APQ$ and AB are in the same vertical plane.

The following method can then be employed.

In Fig. 83 let AB represent the height to be measured.

Taking a point P, measure a horizontal distance PQ in any suitable direction.

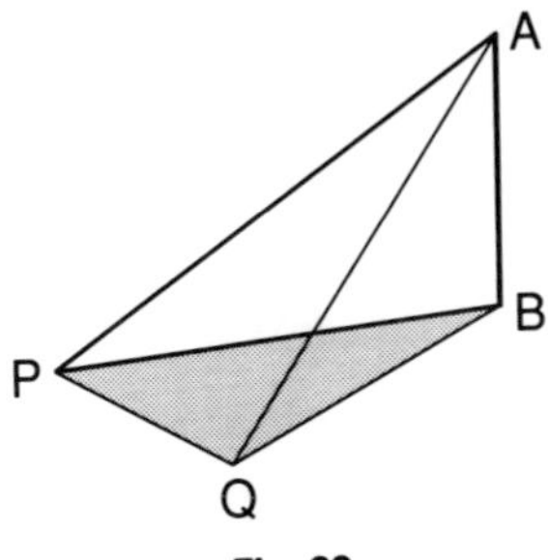


**Fig. 83.**

At P measure

(1) $\angle APB$, the angle of elevation of A,
(2) $\angle APQ$, the bearing of Q from A taken at P.

At Q measure $\angle AQP$, the bearing of P from Q, taken at Q.
Then in $\triangle APQ$.

PQ is known.
$\angle APQ$ is known.
$\angle AQP$ is known.

$\therefore$ $\triangle APQ$ can be solved as in Case III, of section 104.
Thus AP is found and $\angle APB$ is known.

$$\therefore \quad AB = AP \sin APB$$

As a check $\angle AQB$ can be observed and AQ found as above.

Then $\quad AB = AQ \sin AQB$

It should be noted that the distances PB and QB can be determined if required.

*Alternative method.*

Instead of measuring the angles APQ, AQB, we may, by using a theodolite, measure

$$\angle BPQ \text{ at } P$$

and $$\angle PQB \text{ at } Q$$

Then in $\triangle$PQB.

PQ is known.
$\angle$s BPQ, BQP are known.

$\therefore$ $\triangle$PQB can be solved as in Case III, section 104.
Thus BP is determined.
Then $\angle$APB being known

$$AB = PB \tan APB$$

As a check, AB can be found by using BQ and $\angle$AQB.

## 109 Distance of an inaccessible object

Suppose that A (Fig. 84) is an inaccessible object whose distance is required from an observer at P.

A distance PQ is measured in any suitable direction.

$\angle$APQ, the bearing of A with regard to PQ at P is measured.

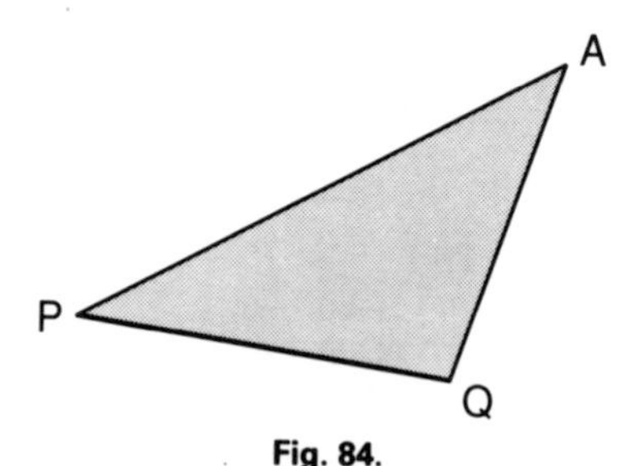


**Fig. 84.**

Also $\angle$AQP, the bearing of A with regard to PQ at Q is measured.

Thus in $\triangle$APQ.

PQ is known.
$\angle$s APQ, AQP are known.

$\therefore$ $\triangle$APQ can be solved as in Case 3, section 104.
Thus AP may be found and, if required, AQ.

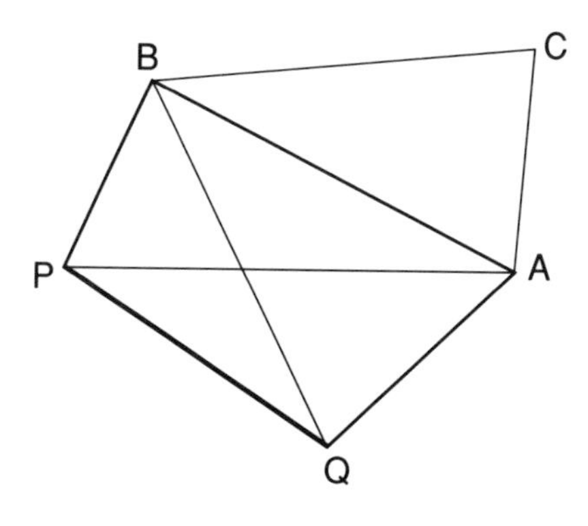


**Fig. 85.**

## 110 Distance between two visible but inaccessible objects

Let A and B (Fig. 85) be two distant inaccessible objects.

Measure any convenient base line PQ.

At P observe $\angle$s APB, BPQ.

At Q observe $\angle$s AQP, AQB.

In $\triangle$APQ.

PQ is known.
$\angle$s APQ, AQP are known.

$\therefore$ $\triangle$ can be solved as in Case III, section 104, and AQ can be found.

Similarly $\triangle$BPQ can be solved and QB can be found.

Then in $\triangle$AQB.

AQ is known.
QB is known.
$\angle$AQB is known.

$\therefore$ $\triangle$AQB can be solved as in Case II, section 103.

Hence AB is found.

A check can be found by solving in a similar manner the $\triangle$APB.

## 111 Triangulation

The methods employed in the last two examples are, in principle, those which are used in triangulation. This is the name given to the method employed in surveying a district, obtaining its area, etc. In practice there are complications such as corrections for sea level and, over large districts, the fact that the earth is approximately a sphere necessitates the use of spherical trigonometry.

Over small areas, however, the error due to considering the surface as a plane, instead of part of a sphere, is, in general, very small, and approximations are obtained more readily than by using spherical trigonometry.

The method employed is as follows:

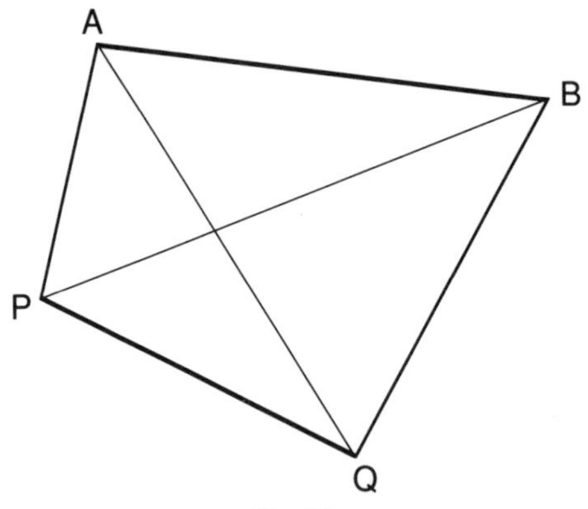


**Fig. 86.**

A measured distance PQ (Fig. 86), called a **base line**, is marked out with very great accuracy on suitable ground. Then a point A is selected and its bearings from P and Q, i.e. ∠s APQ, AQP, are observed. PQ being known, the △APQ can now be solved as in Case III and its area determined.

Next, another point B is selected and the angles BPA, BAP measured.

Hence, as PA has been found from △APQ, △APB can be solved (Case III) and its area found.

Thus the area of the quadrilateral PQAB can be found.

This can be checked by joining BQ.

The △s BPQ, ABQ can now be solved and their areas determined.

Hence we get once more the area of the quadrilateral PQAB.

A new point C can now be chosen.

Using the same methods as before △ABC can be solved.

By repeating this process with other points and a network of triangles a whole district can be covered.

Not only is it essential that the base line should be measured with minute accuracy, but an extremely accurate measurement of the angles is necessary. Checks are used at every stage, such as adding the angles of a triangle to see if their sum is 180°, etc.

The instruments used, especially the theodolite, are provided with verniers and microscopic attachments to secure accurate readings.

As a further check at the end of the work, or at any convenient stage, one of the lines whose length has been found by calculation, founded on previous calculations, can be used as a base line, and the whole survey worked backwards, culminating with the calculation of the original measured base line.

## 112 Worked examples

We will now consider some worked examples illustrating some of the above methods, as well as other problems solved by similar methods.

*Example 1*: Two points lie due W. of a stationary balloon and are 1000 m apart. The angles of elevation at the two points are 21.25° and 18°. Find the height of the balloon.

This is an example of the problem discussed under (b) in section 108.

In Fig. 87

$$\begin{aligned} \angle AQB &= 21.25° \\ \therefore \quad \angle AQP &= 158.75° \\ \angle APQ &= 18° \\ \therefore \quad \angle PAQ &= 3.25° \end{aligned}$$

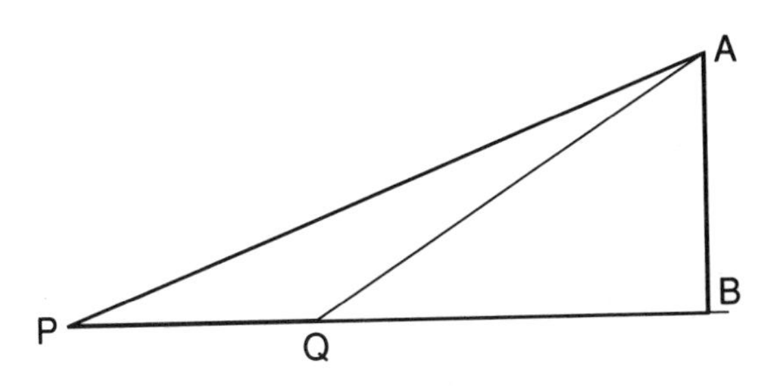


Fig. 87.

$\triangle$APQ is solved as in Case III.

$$\frac{AP}{\sin AQP} = \frac{PQ}{\sin PAQ}$$

$$\therefore \quad \frac{AP}{\sin 158.75^\circ} = \frac{1000}{\sin 3.25^\circ}$$

$$AP = 1000 \times \sin 158.75^\circ \div \sin 3.25^\circ$$
$$= 6393.02$$

whence $\quad AP = 6393$

also $\quad AB = PA \sin 18^\circ$

$$= 6393 \sin 18^\circ$$
$$= 1975.55$$
$$\therefore \quad AB = 1976 \text{ m}$$

*Example 2*: A balloon is observed from two stations A and B at the same horizontal level, A being 1000 m north of B. At a given instant the balloon appears from A to be in a direction N. 32.2° E., and to have an elevation 53.42°, while from B it appears in a direction N. 21.45° E. Find the height of the balloon.

This is an example of (c) above.

In Fig. 88 PQ represents the height of the balloon at P above the ground.

$$\angle NAQ = 33.2^\circ$$
$$\angle ABQ = 21.45^\circ$$
$$\angle PAQ = 53.42^\circ$$

We first solve the $\triangle$ABQ and so find AQ.

$$\angle BAQ = 180^\circ - 33.2^\circ = 146.8^\circ$$
$$\angle AQB = 180^\circ - (BAQ + ABQ)$$
$$= 180^\circ - 168.25^\circ$$
$$= 11.75^\circ$$

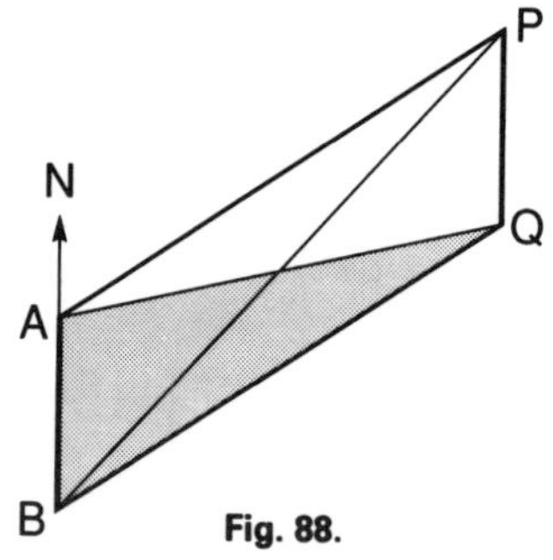


Fig. 88.

The $\triangle ABQ$ can now be solved as in Case 3.

Then $$\frac{AQ}{\sin ABQ} = \frac{AB}{\sin AQB}$$

$$AQ = 1000 \times \sin 21.45° \div \sin 11.75°$$
$$= 1795.75$$

$$\therefore \quad \frac{AQ}{\sin 21.45°} = \frac{1000}{\sin 11.75°}$$

whence $AQ = 1796$

Now $PQ = AQ \tan PAQ$

$\therefore \quad PQ = 1796 \tan 53.42°$

whence $PQ = 2420$ m

*Example 3*: A man who wishes to find the width of a river measures along a level stretch on one bank, a line AB, 150 m long. From A he observes that a post P on the opposite bank is placed so that $\angle PAB = 51.33°$, and $\angle PBA = 69.2°$. What was the width of the river?

In Fig. 89, AB represents the measured distance, 150 m long.
P is the post on the other side of the river.
PQ, drawn perpendicular to AB, represents the width of the river.

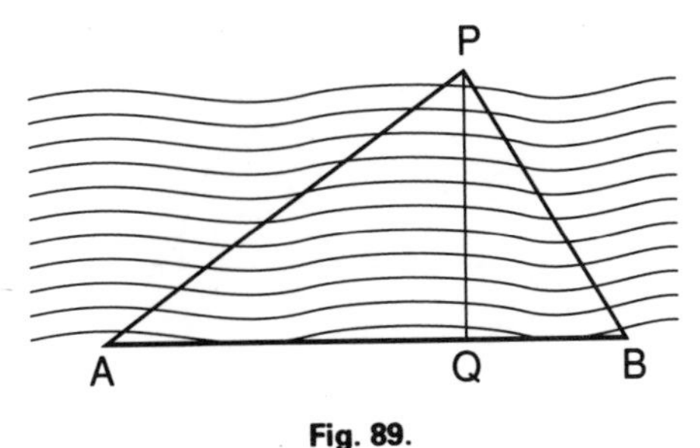


**Fig. 89.**

To find PQ we must first solve the $\triangle APB$.
Then knowing PA or PB we can readily find PQ.
$\triangle APB$ is solved as in Case 3,

$$\angle PAB = 51.33°, \angle PBA = 62.2°$$

$$\therefore \quad \angle APB = 180° - 51.33° + 62.2° = 66.47°$$

$$\frac{PB}{AB} = \frac{\sin 51.33°}{\sin 66.47°}$$

$$\therefore \quad PB = \frac{150 \times \sin 51.33°}{\sin 66.47°}$$

$$\therefore \quad PB = 127.7$$

Again $\quad PQ = PB \sin 66.2°$
whence $\quad PQ = 113$ m

This may be checked by finding PA in △PAB and then finding PQ as above.

*Example 4*: A and B are two ships at sea. P and Q are two stations, 1100 m apart, and approximately on the same horizontal level as A and B. At P, AB subtends an angle of 49° and BQ an angle of 31°. At Q, AB subtends an angle of 60° and AP an angle of 62°. Calculate the distance between the ships.

Fig. 90 represents the given angles and the length PQ (not drawn to scale).

AB can be found by solving either △PAB or △QAB.
To solve △PAB we must obtain AP and BP.
AP can be found by solving △APQ.
BP can be found by solving △PBQ.
In both △s we know one side and two angles.
∴ the △ can be solved as in Case III.

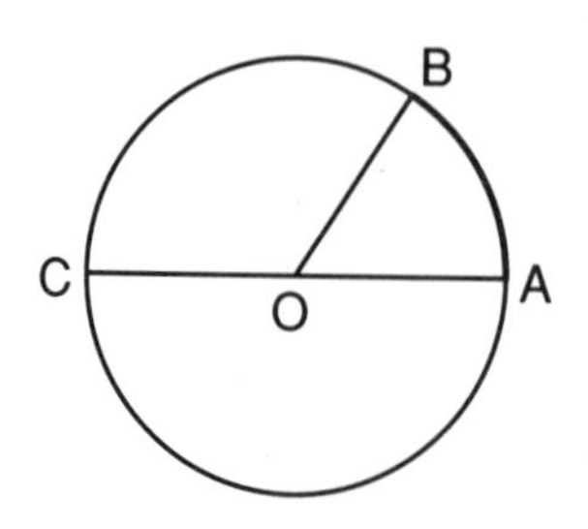


**Fig. 90.**

(1) To solve △APQ and find AP

In △APQ

$$\angle APQ = \angle APB + \angle BPQ$$
$$= 49° + 31° = 80°$$
$$\therefore \quad \angle PAQ = 180° - (80° + 62°) = 38°$$

Using the sine rule $\frac{AP}{PQ} = \frac{\sin 62°}{\sin 38°}$

$$\begin{aligned} AP &= PQ \times \sin 62° \div \sin 38° \\ &= 1100 \times \sin 62° \div \sin 38° \\ &= 1577.56 \\ \therefore \quad AP &= 1578 \text{ m} \end{aligned}$$

(2) To solve $\triangle BPQ$ and find BP

$$\begin{aligned} \angle PQB &= \angle AQB + \angle AQB \\ &= 60° + 62° = 122° \\ \therefore \quad \angle PBQ &= 180° - (31° + 122°) = 27° \end{aligned}$$

Using sine rule $\frac{BP}{PQ} = \frac{\sin 122°}{\sin 27°}$

$$\begin{aligned} BP &= PQ \times \sin 122° \div \sin 27° \\ &= 1100 \times \sin 122° \div \sin 27° \\ &= 2054.79 \\ \therefore \quad BP &= 2055 \text{ m} \end{aligned}$$

(3) To solve $\triangle APB$ and find AB

We know $AP = 1578$ (= c say)
$BP = 2055$ (= b)
$\angle APB = 49°$ (= A)

$\therefore$ Solve as in Case II, section 103.

$$\begin{aligned} b &= 2055 \\ c &= 1578 \\ \hline b + c &= 3633 \\ b - c &= 477 \\ B + C &= 180° - 49° \\ &= 131° \end{aligned}$$

Formula used.

$$\tan \frac{B - C}{2} = \frac{b - c}{b + c} \cot \frac{A}{-}$$

Substituting

$$\begin{aligned} \tan \frac{B - C}{2} &= \frac{477}{3633} \cot 24.5° \\ &= 477 \div 3633 \div \tan 24.5° \\ &= 0.2881 \\ &= \tan 16.07° \end{aligned}$$

$$\therefore \quad \frac{B - C}{2} = 16.07°$$

$$B - C = 32.14°$$

Also

$$\begin{aligned} B + C &= 131° \\ \therefore \quad 2B &= 163.14° \\ B &= 81.57° \\ 2C &= 98.86° \\ C &= 49.43° \\ \therefore \quad \angle PAB &= 81.57° \\ \angle PBA &= 49.86° \end{aligned}$$

(4) To find AB use the sine rule

$$\frac{AB}{AP} = \frac{\sin 49°}{\sin 49.43°}$$

$$\therefore \quad AB = \frac{1578 \times \sin 49°}{\sin 49.43°}$$

$$\therefore \quad AB = 1568 \text{ m}$$

This can be checked by solving $\triangle AQB$ and so obtaining AQ and QB.

## *Exercise 20*

**1** A man observes that the angle of elevation of a tree is 32°. He walks 8 m in a direct line towards the tree and then finds that the angle of elevation is 43°. What is the height of the tree?

**2** From a point Q on a horizontal plane the angle of elevation of the top of a distant mountain is 22.3°. At a point P, 500 m further away in a direct horizontal line, the angle of elevation of the mountain is 16.6°. Find the height of the mountain.

**3** Two men stand on opposite sides of a church steeple and in the same straight line with it. They are 1.5 km apart. From one the angle of elevation of the top of the tower is 15.5° and the other 28.67°. Find the height of the steeple in metres.

**4** A man wishes to find the breadth of a river. From a point on one bank he observes the angle of elevation of a high building on the edge of the opposite bank to be 31°. He then walks 110 m away from the river to a point in the same plane as the previous position and the building he has observed. He finds that the angle of elevation of the building is now 20.92°. What was the breadth of the river?

**5** A and B are two points on opposite sides of swampy ground. From a point P outside the swamp it is found that PA is

882 metres and PB is 1008 metres. The angle subtended at P by AB is 55.67°. What was the distance between A and B?

**6** A and B are two points 1.8 km apart on a level piece of ground along the bank of a river. P is a post on the opposite bank. It is found that $\angle PAB = 62°$ and $\angle PBA = 48°$. Find the width of the river.

**7** The angle of elevation of the top of a mountain from the bottom of a tower 180 m high is 26.42°. From the top of the tower the angle of elevation is 25.3°. Find the height of the mountain.

**8** Two observers 5 km apart take the bearing and elevation of a balloon at the same instant. One finds that the bearing is N. 41° E, and the elevation 24°. The other finds that the bearing is N. 32° E, and the elevation 26.62°. Calculate the height of the balloon.

**9** Two landmarks A and B are observed by a man to be at the same instant in a line due east. After he has walked 4.5 km in a direction 30° north of east, A is observed to be due south while B is 38° south of east. Find the distance between A and B.

**10** At a point P in a straight road PQ it is observed that two distant objects A and B are in a straight line making an angle of 35° at P with PQ. At a point C 2 km along the road from P it is observed that $\angle ACP$ is 50° and angle BCQ is 64°. What is the distance between A and B?

**11** An object P is situated 345 m above a level plane. Two persons, A and B, are standing on the plane, A in a direction south-west of P and B due south of P. The angles of elevation of P as observed at A and B are 34° and 26° respectively. Find the distance between A and B.

**12** P and Q are points on a straight coast line, Q being 5.3 km east of P. A ship starting from P steams 4 km in a direction $65\frac{1}{2}°$ N. of E.

Calculate:

(1) The distance the ship is now from the coast-line.
(2) The ship's bearing from Q.
(3) The distance of the ship from Q.

**13** At a point A due south of a chimney stack, the angle of elevation of the stack is 55°. From B due west of A, such that AB = 100 m, the elevation of the stack is 33°. Find the height of the stack and its horizontal distance from A.

**14** AB is a base line 0.5 km long and B is due west of A. At B a point P bears 65.7° north of west. The bearing of P from AB at A is 44.25° N. of W. How far is P from A?

**15** A horizontal bridge over a river is 380 m long. From one end, A, it is observed that the angle of depression of an object, P vertically beneath the bridge, on the surface of the water is 34°. From the other end, B, the angle of depression of the object is 62°. What is the height of the bridge above the water?

**16** A straight line AB, 115 m long, lies in the same horizontal plane as the foot of a church tower PQ. The angle of elevation of the top of the tower at A is 35°. ∠QAB is 62° and ∠QBA is 48°. What is the height of the tower?

**17** A and B are two points 1500 metres apart on a road running due west. A soldier at A observes that the bearing of an enemy's battery is 25.8° north of west, and at B, 31.5° north of west. The range of the guns in the battery is 5 km. How far can the soldier go along the road before he is within range, and what length of the road is within range?

# 10

# Circular Measure

**113** In chapter 1, when methods of measuring angles were considered, a brief reference was made to 'circular measure' (section 6(c)), in which the unit of measurement is an angle of fixed magnitude, and not dependent upon any arbitrary division of a right angle. We now proceed to examine this in more detail.

## 114 Ratio of the circumference of a circle to its diameter

The subject of 'circular measure' frequently causes difficulties. In order to make it as simple as possible we shall assume, without mathematical proof, the following theorem.

*The ratio of the circumference of a circle to its diameter is a fixed one for all circles.*

This may be expressed in the form:

$$\frac{\text{Circumference}}{\text{diameter}} = \text{a constant}$$

It should, of course, be noted that the ratio of the circumference of a circle to its radius is also constant and the value of the constant must be twice that of the circumference to the diameter.

## 115 The value of the constant ratio of circumference to diameter

You can obtain a fair approximation to the value of the constant by various simple experiments. For example, you can wrap a thread round a cylinder – a glass bottle will do – and so obtain the length of the circumference. You can measure the outside diameter by callipers. The ratio of circumference to diameter thus found will probably give a result somewhere about 3.14.

It is also possible to obtain a much more accurate result by the method devided by Archimedes. The perimeter of a regular polygon **inscribed** in a circle can readily be calculated. The perimeter of a corresponding **escribed** polygon can also be obtained. The mean of these two results will give an approximation to the ratio. By increading the number of sides a still more accurate value can be obtained.

This constant is denoted by the Greek letter $\pi$ (pronounced 'pie').

Hence since $$\frac{\text{circumference}}{\text{diameter}} = \pi$$

$$\therefore \quad \text{circumference} = \pi \times \text{diameter}$$

$$\text{or} \qquad \mathbf{c = 2\pi r}$$

where $c$ = circumference and $r$ = radius.

By methods of advanced mathematics $\pi$ can be calculated to any required degree of accuracy.

To seven places

$$\pi = 3.1415927$$

For many purposes we take

$$\pi = 3.1416$$

which is roughly $\pi = \frac{22}{7}$

It is not possible to find any arithmetical fraction which exactly represents the value of $\pi$. Hence $\pi$ is called 'incommensurable'.

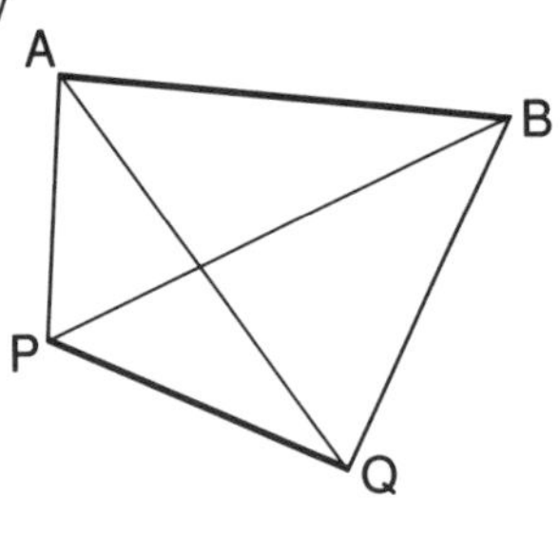


**Fig. 91.**

## 116 The unit of circular measure

As has been stated in section 6(c) the unit of circular measure is the angle subtended at the centre of a circle by an arc whose length is equal to that of the radius.

Thus in Fig. 91 the **length** of the arc AB is equal to r, the radius of the circle. The angle AOB is the unit by which angles are measured, and is termed a **radian**.

*Definition. A radian is the angle subtended at the centre of a circle by an arc equal in length to the radius.*

Note that since
the **circumference** is $\pi$ times the **diameter**
the **semicircular arc** is $\pi$ times the **radius**
or arc of semicircle $= \pi r$.

By Theorem 17, section 18.
*The angles at the centre of a circle are proportional to the arcs on which they stand.*

Now in Fig. 91 the arc of the semicircle ABC subtends two right angles, and the arc AB subtends 1 radian and as semicircle arc is $\pi$ times arc AB

$\therefore$ angle subtended by the semicircular arc is $\pi$ times the angle subtended by arc AB.

i.e. $\qquad$ 2 right angles $= \pi$ radians

or $\qquad 180° = \pi$ radians

## 117 The number of degrees in a radian

As shown above $\pi$ radians $= 180°$

$$\therefore \quad 1 \text{ radian} = \frac{180°}{\pi}$$

$$= 57.2958° \text{ (approx.)}$$

$$\therefore \quad 1 \text{ radian} = 57°\ 17'\ 45'' \text{ (approx.)}$$

Most scientific calculators have a facility (possibly a DRG key) which allows you to change angles in radians to degrees and vice-versa. (For more details of this feature, see section 37 on page 36.)

## 118 The circular measure of any angle

In a circle of radius r, Fig. 92, let AOD be any angle and let $\angle$AOB represent a radian.

$\therefore$ length of arc AB $=$ r.

$$\text{Number of radians in } \angle AOD = \frac{\angle AOD}{\angle AOB}$$

$\therefore$ By Theorem 17 quoted above

$$\frac{\angle AOD}{\angle AOB} = \frac{\text{arc AD}}{\text{arc AB}}$$

If $\theta$ = number of radians in $\angle$AOD

then $$\theta = \frac{\text{arc AD}}{r}$$

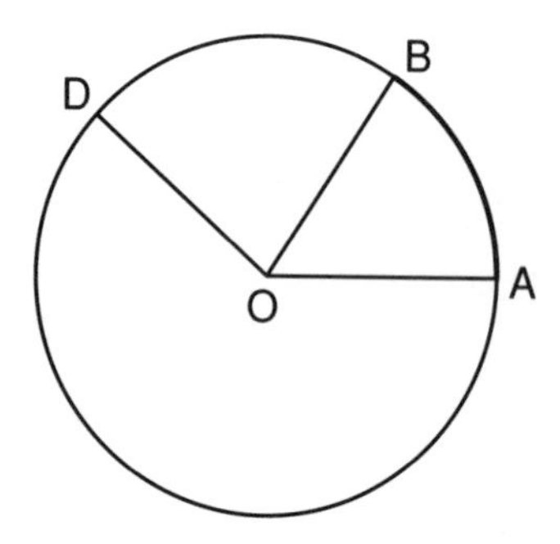


**Fig. 92.**

## 119 To convert degrees to radians

Since $$180° = \pi \text{ radians}$$

$$\therefore \quad 1° = \frac{\pi}{180} \text{ radians}$$

and $$\theta° = \left(\theta \times \frac{\pi}{180}\right) \text{ radians}$$

## 120 To find the length of an arc

Let $\quad a$ = length of arc
and $\quad \theta$ = number of radians in angle

Then as shown in section 118

$\frac{\text{arc}}{\text{radius}}$ = number of radians in the angle the arc subtends.

$$\therefore \quad \frac{a}{r} = \theta \qquad \text{(section 118)}$$

and $$\mathbf{a = r\theta}$$

**121** In more advanced mathematics, circular measure is always employed except in cases in which, for practical purposes, we need to use degrees. Consequently when we speak of an angle $\theta$, it is generally understood that we are speaking of $\theta$ radians. Thus when referring to $\pi$ radians, the equivalent of two right angles, we commonly speak of the angle $\pi$. Hence we have the double use of the symbol:

(1) as the constant ratio of the circumference of a circle to its diameter;
(2) as short for $\pi$ radians, i.e. the equivalent of 180°.

In accordance with this use of $\pi$, angles are frequently expressed as multiples or fractions of it.

Thus

$$2\pi = 360°$$

$$\frac{\pi}{2} = 90°$$

$$\frac{\pi}{4} = 45°$$

$$\frac{\pi}{3} = 60°$$

$$\frac{\pi}{6} = 30°$$

$\pi$ is not usually evaluated in such cases, except for some special purpose.

## *Exercise 21*

**1** What is the number of degrees in each of the following angles expressed in radians: $\frac{\pi}{3'}, \frac{\pi}{12'}, \frac{3\pi}{2'}, \frac{2\pi}{3'}, \frac{3\pi}{4'}$?

**2** Write down from the tables the following ratios:

(*a*) $\sin \frac{\pi}{5}$ (*b*) $\cos \frac{\pi}{8}$ (*c*) $\sin \frac{\pi}{10}$

(*d*) $\cos \frac{3\pi}{8}$ (*e*) $\sin \left(\frac{\pi}{3} + \frac{\pi}{4}\right)$

**3** Express in radians the angles subtended by the following arcs:

(*a*) arc = 11.4 mm, radius = 2.4 mm
(*b*) arc = 5.6 cm, radius = 2.2 cm

**4** Express the following angles in degrees and minutes:

(*a*) 0.234 radian (*b*) 1.56 radian

**5** Express the following angles in radians, using fractions of $\pi$:

(*a*) 15° (*b*) 72° (*c*) 66° (*d*) 105°

**6** Find the length of the arc in each of the following cases:

(1) $r = 2.3$ cm, $\theta = 2.54$ radians
(2) $r = 12.5$ m, $\theta = 1.4$ radians

**7** A circular arc is 154 cm long and the radius of the arc is 252 cm. What is the angle subtended at the centre of the circle, in radians and degrees?

**8** Express a right angle in radians, not using a multiple of $\pi$.

**9** The angles of a triangle are in the ratio of 3:4:5. Express them in radians.

# 11

# Trigonometrical Ratios of Angles of any Magnitude

**122** In chapter 3 we dealt with the trigonometrical ratios of acute angles, i.e. angles in the first quadrant. In chapter 5 the definitions of these ratios were extended to obtuse angles, or angles in the second quadrant. But in mathematics we generalise and consequently in this chapter we proceed to consider the ratios of angles of any magnitude.

In section 5, chapter 1, an angle was defined by the rotation of a straight line from a fixed position and round a fixed centre, and there was no limitation as to the amount of rotation. The rotating line may describe any angle up to 360° or one complete rotation, and may then proceed to two, three, four – to any number of complete rotations in addition to the rotation made initially.

## 123 Angles in the third and fourth quadrants

We will first deal with angles in the third and fourth quadrants, and thus include all those angles which are less than 360° or a complete rotation.

Before proceeding to the work which follows you are advised to revise section 68, in chapter 5, dealing with positive and negative lines.

In section 70 it was shown that the ratios of angles in the second quadrant were defined in the same fundamental method as those of angles in the first quadrant, the only difference being that in obtaining the values of the ratios we have to take into consideration the signs of the lines employed, i.e. whether they are positive or negative.

It will now be seen that, with the same attention to the signs of the lines, the same definitions of the trigonometrical ratios will apply, whatever the quadrant in which the angle occurs.

In Fig. 93 there are shown in separate diagrams, angles in the four quadrants. In each case from a point P on the rotating line a perpendicular PQ is drawn to the fixed line OX, produced in the cases of the second and third quadrants.

Thus we have formed, in each case, a triangle OPQ, using the sides of which we obtain, in each quadrant, the ratios as follows, denoting $\angle AOP$ by $\theta$.

Then, in each quadrant

$$\sin\theta = \frac{PQ}{OP}$$

$$\cos\theta = \frac{OQ}{OP}$$

$$\tan\theta = \frac{PQ}{OQ}$$

We now consider the signs of these lines in each quadrant.

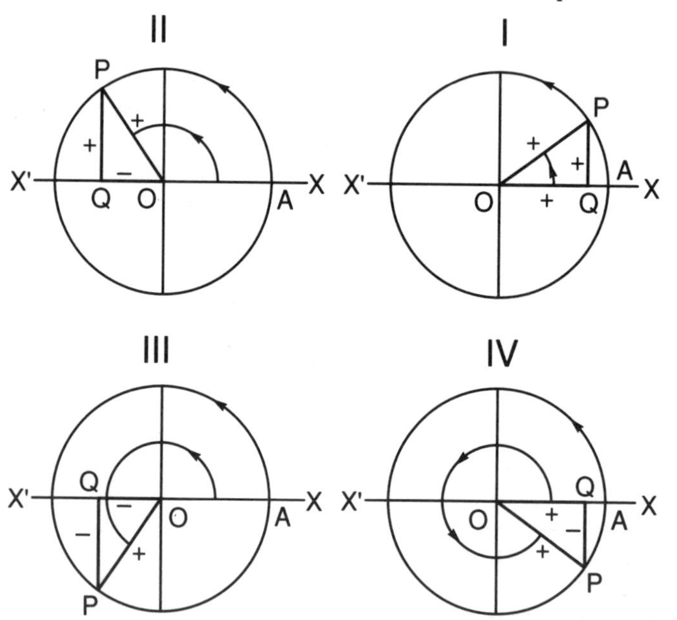


**Fig. 93.**

(1) In the first quadrant.

All the lines are +ve.

∴ All the ratios are +ve.

(2) In the second quadrant.

OQ is −ve

∴ sin $\theta$ is +ve

cos $\theta$ is −ve

tan $\theta$ is −ve

(3) In the third quadrant.

OQ and PQ are −ve

∴ sin $\theta$ is −ve

cos $\theta$ is −ve

tan $\theta$ is +ve

(4) In the fourth quadrant.

PQ is −ve

∴ sin $\theta$ is −ve

cos $\theta$ is +ve

tan $\theta$ is −ve

*Note* The cosecant, secant and tangent will, of course, have the same signs as their reciprocals. These results may be summarised as follows:

| | Quadrant II | Quadrant I | |
|---|---|---|---|
| sine + | sin, +ve<br>cos, −ve<br>tan, −ve | sin, +ve<br>cos, +ve<br>tan, +ve | all + |
| | **Quadrant III** | **Quadrant IV** | |
| tan + | sin, −ve<br>cos, −ve<br>tan, +ve | sin, −ve<br>cos, +ve<br>tan, −ve | cos + |

## 124 Variations in the sine of an angle between 0° and 360°

These have previously been considered for angles in the first and second quadrants. Summarising these for completeness, we will examine the changes in the third and fourth quadrants.

Construct a circle of unit radius (Fig. 94) and centre O. Take on the circumference of this a series of points $P_1$, $P_2$, $P_3$ . . . and draw perpendiculars to the fixed line XOX′. Then the radius being of unit length, these perpendiculars, in the scale in which OA represents unity, will represent the sines of the corresponding angles.

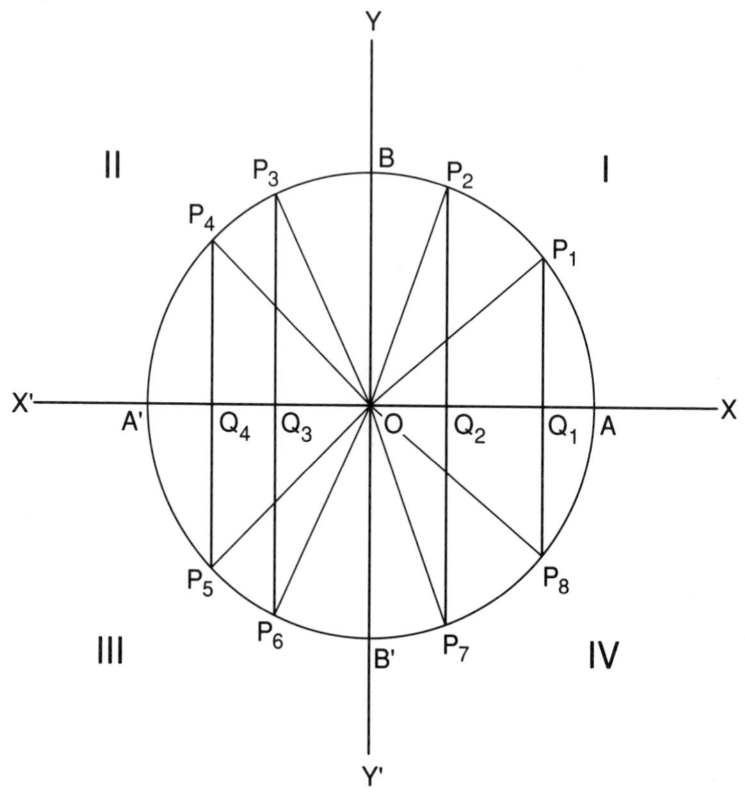


**Fig. 94.**

By observing the changes in the lengths of these perpendiculars we can see, throughout the four quadrants, the changes in the value of the sine.

In quadrant I

sin θ is +ve and increasing from 0 to 1

In quadrant II

sin θ is +ve and decreasing from 1 to 0

In quadrant III

sin θ is −ve

Now the actual lengths of the perpendiculars is increasing, but as they are −ve, the value of the sine is actually decreasing, and at 270° is equal to −1.

∴ *The sine decreases in this quadrant from 0 to −1*

In quadrant IV

sin θ is −ve

The lengths of the perpendiculars are decreasing, but as they are −ve, their values are increasing and at 360° the sine is equal to sin 0° and is therefore zero.

∴ *sin θ is increasing from −1 to 0*

## 125 Graphs of sin θ and cosec θ

By using the values of sines obtained in the method shown above (Fig. 94) or by taking the values of sines from the tables, a graph of the sine between 0° and 360° can now be drawn. It is shown in Fig. 95, together with that of cosec θ (dotted line) the changes in which through the four quadrants can be deduced from those of the sine. You should compare the two graphs, their signs, their maximum and minimum values, etc.

## 126 Variations in the cosine of an angle between 0° and 360°

From Fig. 94 you will see that the distances intercepted on the fixed line by the perpendiculars from $P_1$, $P_2$ . . . , viz. $OQ_1$, $OQ_2$ . . . will represent, in the scale in which OA represents unity, the cosines of the corresponding angles. Examining these we see

(1) In quadrant I

The cosine is +ve and decreases from 1 to 0

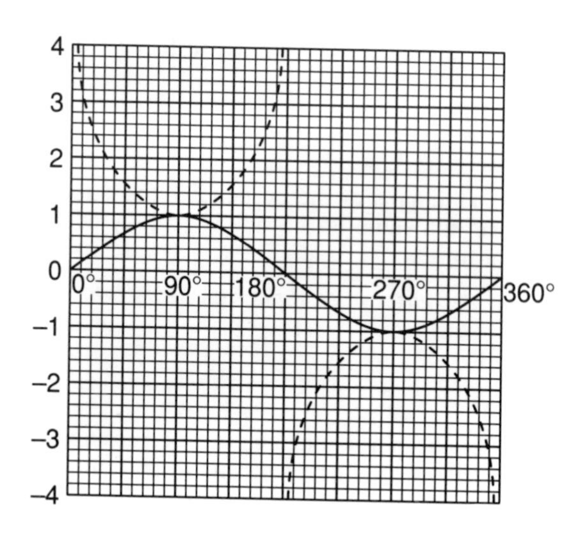


**Fig. 95.**
**Graphs of *sin* θ and *cosec* θ.**
**(*Cosec* θ is dotted.)**

(2) In quadrant II

The cosine is always $-$ve and decreases from 0 to $-1$

(3) In quadrant III

The cosine is $-$ve and always increasing from $-1$ to 0 and $\cos 270° = 0$

(4) In quadrant IV

The cosine is $+$ve and always increasing from 0 to $+1$ since $\cos 360° = \cos 0° = 1$

## 127 Graphs of cos θ and sec θ

In Fig. 96 is shown the graph of cos θ, which can be drawn as directed for the sine in section 125. The curve of its reciprocal, sec θ, is also shown by the dotted curve. Compare these two curves.

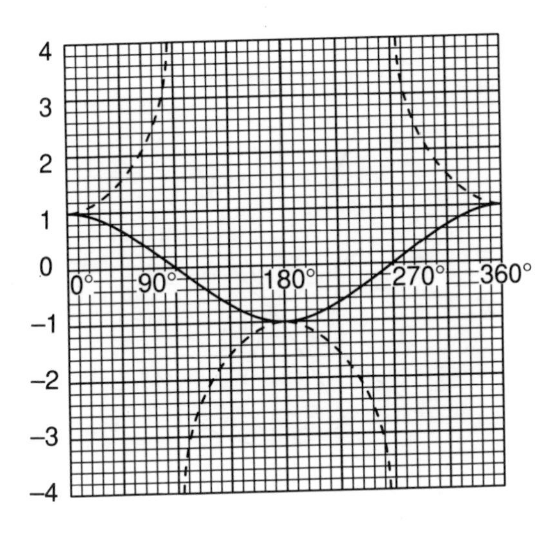


**Fig. 96.**
**Graphs of *cos* θ and *sec* θ (dotted curve).**

## 128 Variations in the tangent between 0° and 360°

The changes in the value of tan θ between 0° and 360° can be seen in Fig. 97, which is an extension of Fig. 39.

The circle is drawn with unit radius.

From A and A′ tangents are drawn to the circle and at right angles to XOX′.

Considering any angle such as $AOP_1$,

$$\tan AOP_1 = \frac{P_1A}{OA} = \frac{P_1A}{1} = P_1A$$

Consequently $P_1A$, $P_2A$, $P_3A'$, $P_4A'$ . . . represent the **numerical** value of the tangent of the corresponding angle.

But account must be taken of the sign.

In quadrants II and III, the denominator of the ratio is $-1$ in numerical value, while in quadrants 3 and 4 the numerator of the fraction is $-$ve.

Consequently the tangent is $+$ve in quadrants 1 and 3 and $-$ve in quadrants II and IV.

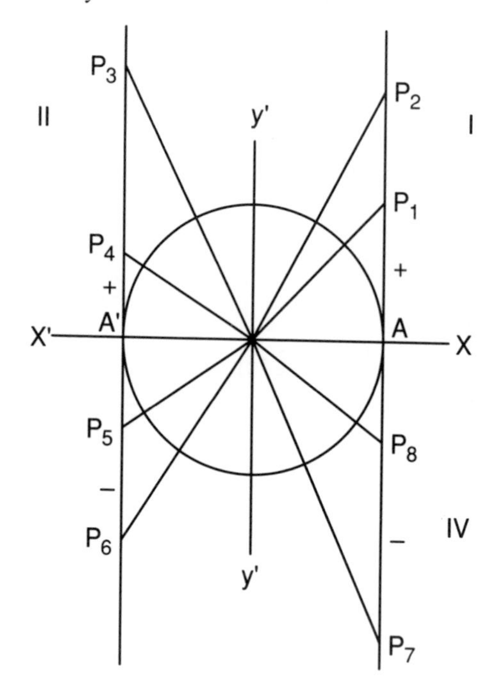


**Fig. 97.**

Considering a particular angle, viz. the $\angle A'OP_5$ in quadrant 3

$$\tan A'OP_5 = \frac{P_5A'}{-OA'}$$

$\therefore$ $\tan\theta$ is +ve and is represented numerically by $P_5A'$.

From such observations of the varying values of $\tan\theta$ the changes between 0° and 360° can be determined as follows:

(1) In quadrant I
 $\tan\theta$ is always +ve and increasing
 It is 0 at 0° and $\rightarrow \infty$ at 90°

(2) In quadrant II
 $\tan\theta$ is always −ve and increasing from $-\infty$ at 90° to 0 at 180°

*Note* When $\theta$ has increased an infinitely small amount above 90°, the tangent becomes −ve.

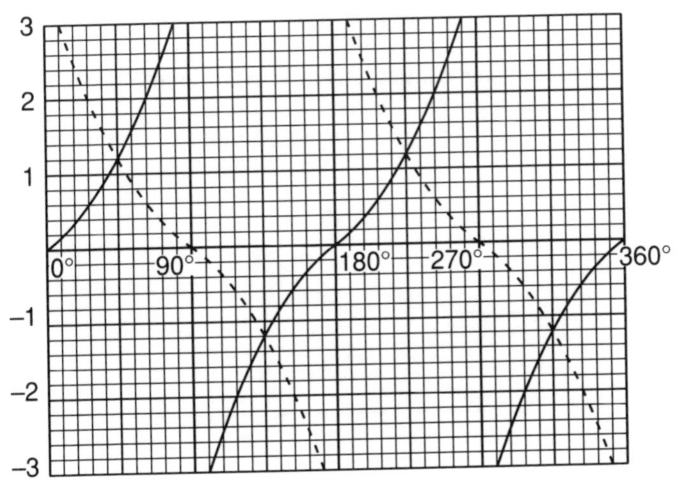


**Fig. 98.**
**Graph of *tan* θ and *cot* θ (dotted line)**

(3) In quadrant III
tan θ is always +ve and increasing
At 180° the tangent is 0 and at 270° tan $\theta \to \infty$

(4) In quadrant IV
tan θ is always −ve and increasing from $-\infty$ at 270° to 0 at 360°

## 129 Graphs of tan θ and cot θ

In Fig. 98 are shown the graphs of tan θ and cot θ (dotted curve) for values of angles between 0° and 360°.

## 130 Ratios of angles greater than 360°

Let ∠AOP (Fig. 99) be any angle, θ, which has been formed by rotation in an anti-clockwise or positive direction from the position OA.

Suppose now that the rotating line continues to rotate in the same direction for a complete rotation or 360° from OP so that it arrives in the same position, OP, as before. The total amount of rotation from OA is now 360° + θ or $(2\pi + \theta)$ radians.

Clearly the trigonometrical ratios of this new angle $2\pi + \theta$ must be the same as $\theta$, so that $\sin(2\pi + \theta) = \sin\theta$, and so for the other ratios.

Similarly if further complete rotations are made so that angles were formed such as $4\pi + \theta$, $6\pi + \theta$, etc., it is evident that the trigonometrical ratios of these angles will be the same at those of $\theta$.

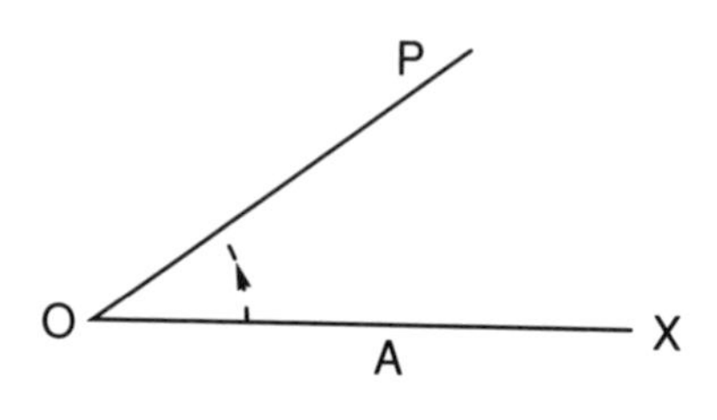


**Fig. 99.**

Turning again to Fig. 99 it is also evident that if a complete rotation is made in a clockwise, i.e. negative, direction, from the position OP, we should have the angle $-2\pi + \theta$. The trigonometrical ratios of this angle, and also such angles as $-4\pi + \theta$, $-6\pi + \theta$, will be the same as those of $\theta$.

All such angles can be included in the general formula

$$2n\pi + \theta$$

where n is any integer, positive or negative.

Referring to the graphs of the ratios in Figs. 95, 96 and 98, it is clear that when the angle is increased by successive complete rotations, the curves as shown, will be repeated either in a positive or a negative direction, and this can be done to an infinite extent.

Each of the ratios is called a **periodic function** of the angle, because the values of the ratio are repeated at intervals of $2\pi$ radians or 360°, which is called the period of the function.

## 131 Trigonometrical ratio of $-\theta$

In Fig. 100 let the rotating line OA rotate in a clockwise, i.e. negative, direction to form the angle AOP. This will be a negative angle. Let it be represented by $-\theta$.

Let the angle AOP′ be formed by rotation in an anti-clockwise i.e. +ve direction and let it be equal to $\theta$.

Then the straight line P′MP completes two triangles.

OMP and OMP′

These triangles are congruent (Theorem 7, section 13) and the angles OMP, OMP′ are equal and ∴ right angles.

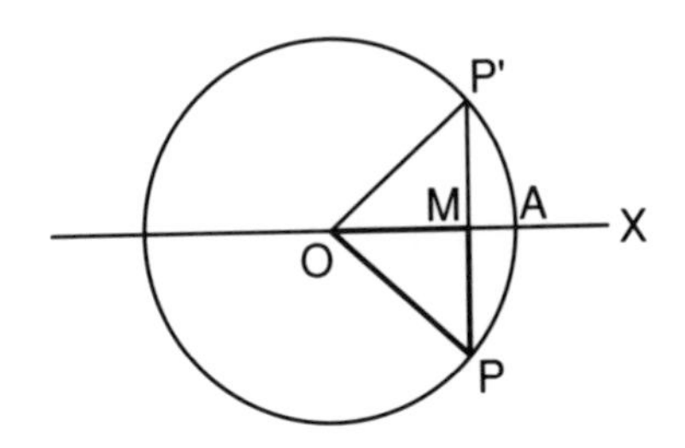


**Fig. 100.**

Then $$\sin(-\theta) = \frac{PM}{OP} = -\frac{P'M}{OP}$$

but $$\frac{PM}{OP} = \sin\theta$$

$$\therefore \quad \sin(-\theta) = -\sin\theta$$

Similarly $$\cos(-\theta) = \frac{OM}{OP} = \frac{OM}{OP'} = \cos\theta$$

Similarly $$\tan(-\theta) = -\tan\theta$$

Collecting these results,

$$\sin(-\theta) = -\sin\theta$$
$$\cos(-\theta) = \cos\theta$$
$$\tan(-\theta) = -\tan\theta$$

From these results you will be able to construct the curves of $\sin\theta$, $\cos\theta$ and $\tan\theta$ for −ve angles. You will see that the curves for −ve angles will be repeated in the opposite direction.

## 132 To compare the trigonometrical ratios of θ and 180° + θ

*Note* If θ is an acute angle, then $180° + \theta$ or $\pi + \theta$ is an angle in the third quadrant.

In Fig. 101 with the usual construction let $\angle$POQ be any acute angle, $\theta$.

Let PO be produced to meet the circle again in P′.

Draw PQ and P′Q′ perpendicular to XOX′.

Then $\angle P'OQ' = \angle POQ = \theta$ (Theorem 1, section 8)

and $\angle AOP' = 180° + \theta$

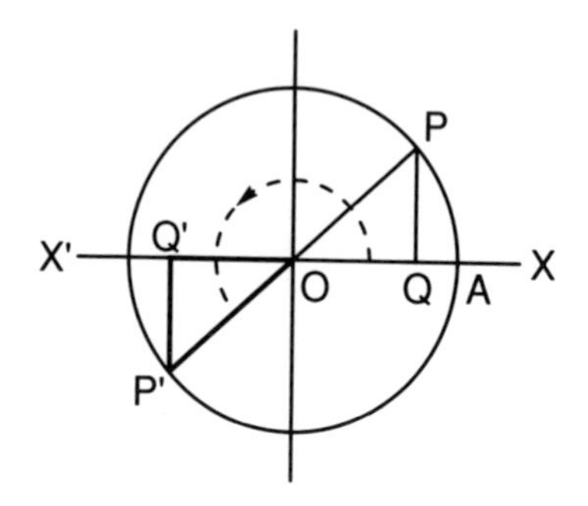


**Fig. 101.**

The $\triangle$s POQ, P′OQ′ are congruent

$$P'Q' = -PQ$$

and

$$OQ' = -OQ$$

Now

$$\sin\theta = \frac{PQ}{OP}$$

and

$$\sin(180° + \theta) = \sin AOP'$$

$$= \frac{P'Q'}{OP'} = \frac{-PQ}{OP} = -\sin\theta$$

$$\therefore \quad \sin\theta = -\sin(180° + \theta)$$

similarly

$$\cos\theta = -\cos(180° + \theta)$$

and

$$\tan\theta = \tan(180° + \theta)$$

## 133 To compare the ratios of $\theta$ and $360° - \theta$

*Note* If $\theta$ is an acute angle, then $360° - \theta$ is an angle in the fourth quadrant.

In Fig. 102 if the acute angle AOP represents $\theta$ then the re-entrant angle AOP, shown by the dotted line represents $360° - \theta$.

The trigonometrical ratios of this angle may be obtained from the sides of the $\triangle$OMP in the usual way and will be the same as those of $-\theta$ (see section 131).

$\therefore$ using the results of section 131 we have

$$\sin (360° - \theta) = -\sin \theta$$
$$\cos (360° - \theta) = \cos \theta$$
$$\tan (360° - \theta) = -\tan \theta$$

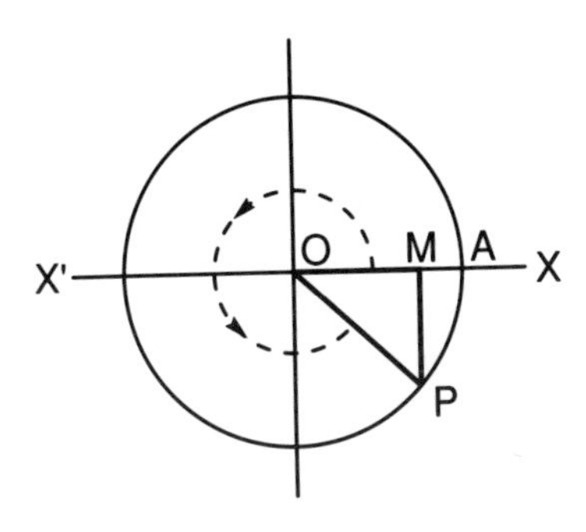


**Fig. 102.**

**134** It will be convenient for future reference to collect some of the results obtained in this chapter, as follows:

$$\sin \theta = \sin (\pi - \theta) = -\sin (\pi + \theta) = -\sin (2\pi - \theta) = -\sin (-\theta)$$

$$\cos \theta = -\cos (\pi - \theta) = -\cos (\pi + \theta) = \cos (2\pi - \theta) = \cos (-\theta)$$

$$\tan \theta = -\tan (\pi - \theta) = \tan (\pi + \theta) = -\tan (2\pi - \theta) = -\tan (-\theta)$$

**135** It is now possible, by use of the above results and using the tables of ratios for acute angles, to write down the trigonometrical ratios of angles of any magnitude.

A few examples are given to illustrate the method to be employed.

*Example 1*: Find the value of sin 245°

We first note that this angle is in the third quadrant

$\therefore$ its sine must be negative.

Next, by using the form of $(180° + \theta)$

$$\sin 245° = \sin (180° + 65°)$$

Thus we can use the appropriate formula of section 134, viz.

$$\sin \theta = \sin (\pi + \theta)$$

Consequently

$$\begin{aligned} \sin (180° + 65°) &= -\sin 65° \\ &= -0.9063 \end{aligned}$$

*Example 2*: Find the value of cos 325°

This angle is in the fourth quadrant and so we use the formulae for values of $360° - \theta$ (see section 133).

In this quadrant the cosine is always +*ve*

$$\begin{aligned} \cos 325° &= \cos 35° \qquad \text{(section 133)} \\ &= 0.8192 \end{aligned}$$

*Example 3*: Find the value of tan 392°

This angle is greater then 360° or one whole revolution.

$$\begin{aligned} \therefore \quad \tan 392° &= \tan (360° + 32°) \\ &= \tan 32° \\ &= 0.6249 \end{aligned}$$

*Example 4*: Find the value of sec 253°

This angle is in the third quadrant.

$\therefore$ we use the formula connected with $(\pi + \theta)$ (see section 132).

Also in this quadrant the cosine, the reciprocal of the secant is −ve.

$$\begin{aligned} \sec 253° &= \sec (180° + 73°) \\ &= -\sec 73° \\ &= -3.4203 \end{aligned}$$

Now use your calculator to check the above results.

*Exercise 22*

**1** Find the sine, cosine and tangent of each of the following angles:

(*a*) 257° (*b*) 201.22°
(*c*) 315.33° (*d*) 343.13°

**2** Find the values of:

(*a*) sin (−51°) (*b*) cos (−42°)
(*c*) sin (−138°) (*d*) cos (−256°)

**3** Find the values of:

(*a*) cosec 251° (*b*) sec 300°
(*c*) cot 321° (*d*) sec 235°)

**4** Find the values of:

(*a*) sin ($\pi$ + 57°) (*b*) cos ($2\pi$ − 42°)
(*c*) tan ($2\pi$ + 52°) (*d*) sin ($4\pi$ + 36°)

## 136 To find the angles which have given trigonometrical ratios

(a) To find all the angles which have a given sine (or cosecant).

We have already seen in section 73 that corresponding to a given sine there are two angles, $\theta$ and 180° − $\theta$, where $\theta$ is the acute angle whose sine is given in the tables. Having now considered angles of any magnitude it becomes necessary to discover what other angles have the given sine.

An examination of the graph of sin $\theta$ in Fig. 95 shows that only two of the angles less than 360° have a given sine, whether it be positive or negative, the two already mentioned above is the sine is +ve, and two in the third and fourth quadrants if it is −ve.

But the curve may extend to an indefinite extent for angles greater than 360°, and for negative angles, and every section corresponding to each additional 360°, positive or negative, will be similar to that shown. Therefore it follows that there will be an infinite number of other angles, two in each section which have the given sine. These will occur at intervals of $2\pi$ radians from those in the first quadrant. There will thus be two sets of such angles.

(1) $\theta, 2\pi + \theta, 4\pi + \theta, \ldots$
(2) $\pi - \theta, 3\pi - \theta, 5\pi - \theta, \ldots$

These two sets include all the angles which have the given sines. They can be summarised as follows:

(1) (any even multiple of $\pi$) + $\theta$
(2) (any odd multiple of $\pi$) − $\theta$

These can be combined together in one formula as follows:

Let n be any integer, positive or negative.
Then sets (1) and (2) are contained in

$$n\pi + (-1)^n \theta$$

The introduction of $(-1)^n$ is a device which ensures that when n is **even**, i.e. we have an even multiple of $\pi$, $(-1)^n = 1$ and the formula becomes $n\pi + \theta$. When n is odd $(-1)^n = -1$ and the formula becomes $n\pi - \theta$.

∴ the general formula for all angles which have a given sine is

$$n\pi + (-1)^n \theta$$

where n is any integer +ve or −ve, and $\theta$ is the smallest angle having the given sine.

*The same formula will clearly hold also for the cosecant.*

(b) To find all the angles which have a given cosine (or secant).

Examining the graph of cos $\theta$ (Fig. 96), it is seen that there are two angles between 0° and 360° which have a given cosine which is +ve, one in the first quadrant and one in the fourth. If the given cosine is −ve, the two angles lie in the second and third quadrants. These two angles are expressed by $\theta$ and $360° - \theta$.

or $\theta$ and $2\pi - \theta$ in radians

As in the case of the sine for angles greater than 360° or for negative angles, there will be two angles with the given sine in the section corresponding to each additional 360°.

There will therefore be two sets:

(1) $\theta, 2\pi + \theta, 4\pi + \theta, \ldots$
(2) $2\pi - \theta, 4\pi - \theta, 6\pi - \theta, \ldots$

These can be combined in one set, viz.:

$$(\text{any even multiple of } \pi) \pm \theta$$

or if n is any integer, positive or negative, this can be expressed by

$$2n\pi \pm \theta$$

$\therefore$ the general formula for all angles with a given cosine is:

$$2n\pi \pm \theta$$

*The formula for the secant will be the same.*

(c) To find all the angles which have a given tangent (or cotangent).

An examination of the graph of tan θ (Fig. 98), shows that there are two angles less than 360° which have the same tangent, viz.:

$$\theta \text{ and } 180^\circ + \theta$$

or

$$\theta \text{ and } \pi + \theta$$

As before, there will be other angles at intervals of $2\pi$ which will have the same tangent. Thus there will be two sets, viz.:

$$\theta,\ 2\pi + \theta,\ 4\pi + \theta,\ \ldots$$
$$\pi + \theta,\ 3\pi + \theta,\ 5\pi + \theta,\ \ldots$$

Combining these it is clear that all are included in the general formula

$$(\text{any multiple of } \pi) + \theta$$

$\therefore$ If n be any integer, positive or negative, the general formula for all angles with a given tangent is

$$n\pi + \theta$$

*The same formula holds for the cotangent.*

Exercises which involve the use of these formulae will occur in the next chapter.

# 12

# Trigonometrical Equations

**137** Trigonometrical equations are those in which the unknown quantities, whose values we require, are the trigonometrical ratios of angles. The angles themselves can be determined when the values of the ratios are known.

The actual form which the answer will take depends on whether we require only the smallest angle corresponding to the ratio, which will be obtained from the tables, or whether we want to include some or all of those other angles which, as we have seen in the previous chapter, have the same ratio.

This can be shown in a very simple example.

*Example*: Solve the equation $2 \cos \theta = 0.842$.

(1) The smallest angle only may be required.

Since $\qquad 2 \cos \theta = 0.842$

$\qquad \cos \theta = 0.421$

From the tables $\qquad \theta = 65.1°$

(2) The angles between 0° and 360° which satisfy the equation may be required.

As we have seen in section 136(b) there is only one other such angle, in the fourth quadrant.

It is given by $2\pi - \theta$ or $360° - \theta$

$\therefore$ This angle $= 360° - 65.1° = 294.9°$

$\therefore$ The two solutions are 65.1° and 294.9°

(3) A general expression for all angles which satisfy the equation may be required.

In this case one of the formulae obtained in the previous chapter will be used.

Thus in section 136(b) all angles with a given cosine are included in the formula

$$2n\pi \pm \theta$$

In this example $\theta = 65.1°$.

$\therefore$ The solution is $2n\pi \pm \cos^{-1} 0.421$

The inverse notation (see section 74) is used to avoid the incongruity of part of the answer $2n\pi$ being in radians, and the other in degrees.

**138** Some of the different types of equations will now be considered.

(a) Equations which involve only one ratio

The example considered in the previous paragraph is the simplest form of this type. Very little manipulation is required unless the equation is quadratic in form.

*Example*: Solve the equation

$$6 \sin^2 \theta - 7 \sin \theta + 21 = 0$$

for values of $\theta$ between 0° and 360°.

Factorising

$$(3 \sin \theta - 2)(2 \sin \theta - 1) = 0$$

whence $\quad 3 \sin \theta - 2 = 0 \quad (1)$

or $\quad 2 \sin \theta - 1 = 0 \quad (2)$

From (1) $\quad \sin \theta = \frac{2}{3} = 0.6667$

$\therefore$ from the tables

$$\theta = 41.82°$$

The only other angle less than 360° with this sine is

$$180° - \theta = 138.18°$$

From (2) $\quad \sin \theta = 0.5$

$$\therefore \quad \theta = 30°$$

and the other angle with this sine is $180° - 30° = 150°$

$\therefore$ the complete solution is

$$41.82°, 138.18°, 30°, 150°$$

*Note* If one of the values of $\sin\theta$ or $\cos\theta$ obtained in an equation is numerically greater than unity, such a root must be discarded as impossible. Similarly values of the secant and cosecant less than unity are impossible solutions from this point of view.

(b) Equations containing more than one ratio of the angle

Manipulation is necessary to replace one of the ratios by its equivalent in terms of the other. To effect this we must use an appropriate formula connected with the ratios such as were proved in chapter 4.

*Example 1*: Obtain a complete solution of the equation

$$3\sin\theta = 2\cos^2\theta.$$

The best plan here is to change $\cos^2\theta$ into its equivalent value of $\sin\theta$. This can be done by the formula

$$\sin^2\theta + \cos^2\theta = 1$$

whence

$$\cos^2\theta = 1 - \sin^2\theta$$

Substituting in the above equation

$$3\sin\theta = 2(1 - \sin^2\theta)$$

Factorising,

$$(2\sin\theta - 1)(\sin\theta + 2) = 0$$

whence

$$\sin\theta + 2 = 0 \qquad (1)$$

or

$$2\sin\theta - 1 = 0 \qquad (2)$$

From (1)

$$\sin\theta = -2$$

This is impossible, and therefore does not provide a solution of the given equation.

From (2)

$$2\sin\theta = 1$$

$$\therefore \sin\theta = 0.5$$

The smallest angle with this sine is $30°$ or $\frac{\pi}{6}$ radians.

Using the general formula for all angles with a given sine, viz.:

$$n\pi + (-1)^n\,\theta$$

The general solution of the equation is

$$\theta = n\pi + (-1)^n \frac{\pi}{6}$$

*Example 2*: Solve the equation

$$\sin 2\theta = \cos^2 \theta$$

giving the values of θ between 0° and 360° which satisfy the equation.

Since $\quad \sin 2\theta = 2 \sin \theta \cos \theta \qquad$ (see section 83)

$\therefore \quad 2 \sin \theta \cos \theta = \cos^2 \theta$

Hence $\quad \cos \theta = 0 \qquad (1)$

or $\quad 2 \sin \theta = \cos \theta \qquad (2)$

From (1) $\quad \theta = 90° \text{ or } 270°$

From (2) $\quad 2 \sin \theta = \cos \theta$

$\therefore \quad 2 \tan \theta = 1$

and $\quad \tan \theta = 0.5$

whence $\quad \theta = 26.57°$

Also $\quad \tan \theta = \tan (180° + \theta) \qquad$ (see section 132)

∴ The other angle less than 360° with this tangent is

$$180° + 26.57°$$
$$= 206.57°$$

∴ The solution is

$$\theta = 26.57° \text{ or } 206.57°$$

∴ The required solution is $\theta = 90°, 270°, 26.57° \text{ or } 206.57°$

## 139 Equations of the form

$$a \cos \theta + b \sin \theta = c$$

where a, b, c are known constants, are important in electrical work and other applications of trigonometry.

This could be solved by using the substitution

$$\cos \theta = \sqrt{1 - \sin^2 \theta}$$

but the introduction of the square root is not satisfactory. We can obtain a solution more readily by the following device.

Since a and b are known it is always possible to find an angle $\alpha$, such that

$$\tan \alpha = \frac{a}{b}$$

as the tangent is capable of having any value (see graph, Fig. 98).

Let ABC (Fig. 103) be a right-angled $\triangle$ in which the sides containing the right angle are a and b units in length.

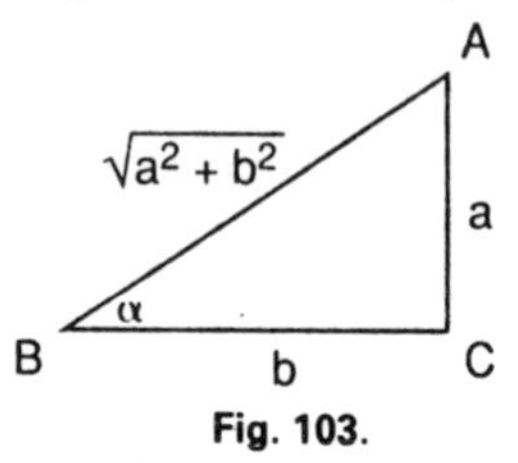


Fig. 103.

Then
$$\tan ABC = \frac{a}{b}$$

$$\therefore \quad \angle ABC = \alpha$$

By the Theorem of Pythagoras:

$$AB = \sqrt{a^2 + b^2}$$

and
$$\frac{a}{\sqrt{a^2 + b^2}} = \sin \alpha$$

$$\frac{b}{\sqrt{a^2 + b^2}} = \cos \alpha$$

$\therefore$ in the equation

$$a \cos \theta + b \sin \theta = c$$

Divide throughout by $\sqrt{a^2 + b^2}$

$$\therefore \quad \frac{a}{\sqrt{a^2 + b^2}} \cos \alpha + \frac{b}{\sqrt{a^2 + b^2}} \sin \alpha = \frac{c}{\sqrt{a^2 + b^2}}$$

$$\therefore \quad \sin \alpha \cos \theta + \cos \alpha \sin \theta = \frac{c}{\sqrt{a^2 + b^2}}$$

$$\therefore \quad \sin (\theta + \alpha) = \frac{c}{\sqrt{a^2 + b^2}}$$

(see section 80, No. 1)

Now $\frac{c}{\sqrt{a^2 + b^2}}$ can be evaluated, since a, b, c are known and provided it is less than unity it is the sine of some angle, say β.

$$\therefore \quad \theta + \alpha = \beta$$

and $$\theta = \beta - \alpha$$

Thus the least value of θ is determined.

*Example* Solve the equation $3 \cos \theta + 4 \sin \theta = 3.5$

In this case $$a = 3, \ b = 4,$$

$$\therefore \quad \sqrt{a^2 + b^2} = \sqrt{9 + 16} = 5$$

Thus $\tan \alpha = \frac{3}{4}$, $\sin \alpha = \frac{3}{5}$, $\cos \alpha = \frac{4}{5}$ and $\alpha = 36.87°$ (from the tables).

∴ Dividing the given equation by 5

$$\tfrac{3}{5} \cos \theta + \tfrac{4}{5} \sin \theta = \frac{3.5}{5}$$

$$\therefore \quad \sin \alpha \cos \theta + \cos \alpha \sin \theta = 0.7$$

$$\therefore \quad \sin (\theta + \alpha) = 0.7$$

But the angle whose sine is 0.7 is 44.42°

$$\therefore \quad \theta + \alpha = 44.42°$$

or $$\theta + 36.87° = 44.42°$$

$$\therefore \quad \theta = 44.42° - 36.87°$$

$$= 7.55°$$

## 139 Variations of a cos θ + b sin θ

This expression is an important one in its application, and the graphical representations of its variation may have to be studied by some students. The variations of the expression may be best studied by using, in a modified form, the device employed above.

By means of the reasoning given in the previous paragraph, the expression can be written in the form

$$\sqrt{a^2 + b^2} \ \{\sin (\theta + \alpha)\}$$

By assigning different values to θ, the only variable in the expression, the variations can be studied and a graph constructed.

## *Exercise 23*

**1** Find the angles less than 360° which satisfy the following equations:

(1) $\sin \theta = 0.8910$ (2) $\cos \theta = 0.4179$
(3) $2 \tan \theta = 0.7$ (4) $\sec \theta = 2.375$

**2** Find the angles less than 360° which satisfy the following equations:

(1) $4 \cos 2\theta - 3 = 0$ (2) $3 \sin 2\theta = 1.8$

**3** Find the angles less than 360° which satisfy the following equations:

(1) $6 \sin \theta = \tan \theta$ (2) $4 \cos \theta = 3 \tan \theta$
(3) $3 \cos^2 \theta + 5 \sin^2 \theta = 4$ (4) $4 \cos \theta = 3 \sec \theta$

**4** Find the angles less than 360° which satisfy the following equations:

(1) $2 \tan^2 \theta - 3 \tan \theta + 1 = 0$
(2) $5 \tan^2 \theta - \sec^2 \theta = 11$
(3) $4 \sin^2 \theta - 3 \cos \theta = 1.5$
(4) $\sin \theta + \sin^2 \theta = 0$

**5** Find general formulae for the angles which satisfy the following equations:

(1) $2 \cos \theta - 0.6578 = 0$
(2) $\frac{1}{2} \sin 2\theta = 0.3174$
(3) $\cos 2\theta + \sin \theta = 1$
(4) $\tan \theta + \cot^2 \theta = 4$

**6** Find the smallest angles which satisfy the equations:

(1) $\sin \theta + \cos \theta = 1.2$
(2) $\sin \theta - \cos \theta = 0.2$
(3) $2 \cos 2\theta + \sin \theta = 2.1$
(4) $4 \cos \theta + 3 \sin \theta = 5$

# Summary of Trigonometrical Formulae

## 1 Complementary angles

$$\sin \theta = \cos (90° - \theta)$$
$$\cos \theta = \sin (90° - \theta)$$
$$\tan \theta = \cot (90° - \theta)$$

## 2 Supplementary angles

$$\sin \theta = \sin (180° - \theta)$$
$$\cos \theta = -\cos (180° - \theta)$$
$$\tan \theta = -\tan (180° - \theta)$$

## 3 Relations between the ratios

$$\tan \theta = \frac{\sin \theta}{\cos \theta}$$

$$\sin^2 \theta + \cos^2 \theta = 1$$
$$\tan^2 \theta + 1 = \sec^2 \theta$$
$$\cot^2 \theta + 1 = \operatorname{cosec}^2 \theta$$

## 4 Compound angles

$$\sin (A + B) = \sin A \cos B + \cos A \sin B$$
$$\sin (A - B) = \sin A \cos B - \cos A \sin B$$
$$\cos (A + B) = \cos A \cos B - \sin A \sin B$$
$$\cos (A - B) = \cos A \cos B + \sin A \sin B$$

$$\tan (A + B) = \frac{\tan A + \tan B}{1 - \tan A \tan B}$$

$$\tan(A - B) = \frac{\tan A - \tan B}{1 + \tan A \tan B}$$

$$\sin P + \sin Q = 2 \sin \frac{P+Q}{2} \cos \frac{P-Q}{2}$$

$$\sin P - \sin Q = 2 \cos \frac{P+Q}{2} \sin \frac{P-Q}{2}$$

$$\cos P + \cos Q = 2 \cos \frac{P+Q}{2} \cos \frac{P-Q}{2}$$

$$\cos Q - \cos P = 2 \sin \frac{P+Q}{2} \sin \frac{P-Q}{2}$$

## 5 Multiple angles

$$\sin 2\theta = 2 \sin \theta \cos \theta$$

$$\cos 2\theta = \cos^2 \theta - \sin^2 \theta$$

$$= 2 \cos^2 \theta - 1$$

$$= 1 - 2 \sin^2 \theta$$

or

$$\cos^2 \theta = \tfrac{1}{2}(1 + \cos 2\theta)$$

$$\sin^2 \theta = \tfrac{1}{2}(1 - \cos 2\theta)$$

$$\tan 2\theta = \frac{2 \tan \theta}{1 - \tan^2 \theta}$$

## 6 Solutions of a triangle

### Case 1 Three sides known

1 $\cos A = \dfrac{b^2 + c^2 - a^2}{2bc}$ (if a, b, c are small)

2 $\tan \dfrac{A}{2} = \sqrt{\dfrac{(s-b)(s-c)}{s(s-a)}}$ (for use with logs)

$$\sin \frac{A}{2} = \sqrt{\frac{(s-b)(s-c)}{bc}}$$

$$\cos \frac{A}{2} = \sqrt{\frac{s(s-a)}{bc}}$$

$$\sin A = \frac{2}{bc} \sqrt{s(s-a)(s-b)(s-c)}$$

**Case 2 Two sides and contained angle known**

$$\tan \frac{B - C}{2} = \frac{b - c}{b + c} \cot \frac{A}{2}$$

**Case 3 Two angles and a side known**

$$\frac{\sin A}{a} = \frac{\sin B}{b} = \frac{\sin C}{c}$$

## 7 Ratios of angles between 0 and $2\pi$ radians

$$\begin{aligned}
\sin \theta &= \sin(\pi - \theta) = -\sin(\pi + \theta) = -\sin(2\pi - \theta) \\
\cos &= -\cos(\pi - \theta) = -\cos(\pi + \theta) = \cos(2\pi - \theta) \\
\tan &= -\tan(\pi - \theta) = \tan(\pi + \theta) = -\tan(2\pi - \theta)
\end{aligned}$$

## 8 Ratios of $\theta$ and $-\theta$

$$\begin{aligned}
\sin \theta &= -\sin(-\theta) \\
\cos \theta &= \cos(-\theta) \\
\tan \theta &= -\tan(-\theta)
\end{aligned}$$

## 9 General formulae for angles with the same ratios as $\theta$

| | |
|---|---|
| sine | $n\pi + (-1)^n \theta$ |
| cosine | $2n\pi \pm \theta$ |
| tangent | $n\pi + \theta$ |

## 10 Circular measure

1 radian = 57° 17′ 45″ (approx.) = 57.2958°

**To convert degrees to radians**

$$\theta^\circ = \left(\theta^\circ \times \frac{\pi}{180}\right) \text{ radians}$$

**Length of an arc**

$$a = r\theta \ (\theta \text{ in radians})$$

# NATURAL SINES

Proportional Parts

| | 0′ | 6′ 0.1 | 12′ 0.2 | 18′ 0.3 | 24′ 0.4 | 30′ 0.5 | 36′ 0.6 | 42′ 0.7 | 48′ 0.8 | 54′ 0.9 | 1′ | 2′ | 3′ | 4′ | 5′ |
|---|---|---|---|---|---|---|---|---|---|---|---|---|---|---|---|
| 0° | 0.0000 | .0017 | .0035 | .0052 | .0070 | .0087 | .0105 | .0122 | .0140 | .0157 | 3 | 6 | 9 | 12 | 15 |
| 1 | 0.0175 | .0192 | .0209 | .0227 | .0244 | .0262 | .0279 | .0297 | .0314 | .0332 | 3 | 6 | 9 | 12 | 15 |
| 2 | 0.0349 | .0366 | .0384 | .0401 | .0419 | .0436 | .0454 | .0471 | .0489 | .0506 | 3 | 6 | 9 | 12 | 15 |
| 3 | 0.0523 | .0541 | .0558 | .0576 | .0593 | .0610 | .0628 | .0645 | .0663 | .0680 | 3 | 6 | 9 | 12 | 15 |
| 4 | 0.0698 | .0715 | .0732 | .0750 | .0767 | .0785 | .0802 | .0819 | .0837 | .0854 | 3 | 6 | 9 | 12 | 14 |
| 5 | 0.0872 | .0889 | .0906 | .0924 | .0941 | .0958 | .0976 | .0993 | .1011 | .1028 | 3 | 6 | 9 | 12 | 14 |
| 6 | 0.1045 | .1063 | .1080 | .1097 | .1115 | .1132 | .1149 | .1167 | .1184 | .1201 | 3 | 6 | 9 | 12 | 14 |
| 7 | 0.1219 | .1236 | .1253 | .1271 | .1288 | .1305 | .1323 | .1340 | .1357 | .1374 | 3 | 6 | 9 | 12 | 14 |
| 8 | 0.1392 | .1409 | .1426 | .1444 | .1461 | .1478 | .1495 | .1513 | .1530 | .1547 | 3 | 6 | 9 | 11 | 14 |
| 9 | 0.1564 | .1582 | .1599 | .1616 | .1633 | .1650 | .1668 | .1685 | .1702 | .1719 | 3 | 6 | 9 | 11 | 14 |
| 10 | 0.1736 | .1754 | .1771 | .1788 | .1805 | .1822 | .1840 | .1857 | .1874 | .1891 | 3 | 6 | 9 | 11 | 14 |
| 11 | 0.1908 | .1925 | .1942 | .1959 | .1977 | .1994 | .2011 | .2028 | .2045 | .2062 | 3 | 6 | 9 | 11 | 14 |
| 12 | 0.2079 | .2096 | .2113 | .2130 | .2147 | .2164 | .2181 | .2198 | .2215 | .2232 | 3 | 6 | 9 | 11 | 14 |
| 13 | 0.2250 | .2267 | .2284 | .2300 | .2317 | .2334 | .2351 | .2368 | .2385 | .2402 | 3 | 6 | 8 | 11 | 14 |
| 14 | 0.2419 | .2436 | .2453 | .2470 | .2487 | .2504 | .2521 | .2538 | .2554 | .2571 | 3 | 6 | 8 | 11 | 14 |
| 15 | 0.2588 | .2605 | .2622 | .2639 | .2656 | .2672 | .2689 | .2706 | .2723 | .2740 | 3 | 6 | 8 | 11 | 14 |
| 16 | 0.2756 | .2773 | .2790 | .2807 | .2823 | .2840 | .2857 | .2874 | .2890 | .2907 | 3 | 6 | 8 | 11 | 14 |
| 17 | 0.2924 | .2940 | .2957 | .2974 | .2990 | .3007 | .3024 | .3040 | .3057 | .3074 | 3 | 6 | 8 | 11 | 14 |
| 18 | 0.3090 | .3107 | .3123 | .3140 | .3156 | .3173 | .3190 | .3206 | .3223 | .3239 | 3 | 6 | 8 | 11 | 14 |
| 19 | 0.3256 | .3272 | .3289 | .3305 | .3322 | .3338 | .3355 | .3371 | .3387 | .3404 | 3 | 5 | 8 | 11 | 14 |
| 20 | 0.3420 | .3437 | .3453 | .3469 | .3486 | .3502 | .3518 | .3535 | .3551 | .3567 | 3 | 5 | 8 | 11 | 14 |
| 21 | 0.3584 | .3600 | .3616 | .3633 | .3649 | .3665 | .3681 | .3697 | .3714 | .3730 | 3 | 5 | 8 | 11 | 14 |
| 22 | 0.3746 | .3762 | .3778 | .3795 | .3811 | .3827 | .3843 | .3859 | .3875 | .3891 | 3 | 5 | 8 | 11 | 13 |
| 23 | 0.3907 | .3923 | .3939 | .3955 | .3971 | .3987 | .4003 | .4019 | .4035 | .4051 | 3 | 5 | 8 | 11 | 13 |
| 24 | 0.4067 | .4083 | .4099 | .4115 | .4131 | .4147 | .4163 | .4179 | .4195 | .4210 | 3 | 5 | 8 | 11 | 13 |
| 25 | 0.4226 | .4242 | .4258 | .4274 | .4289 | .4305 | .4321 | .4337 | .4352 | .4368 | 3 | 5 | 8 | 11 | 13 |
| 26 | 0.4384 | .4399 | .4415 | .4431 | .4446 | .4462 | .4478 | .4493 | .4509 | .4524 | 3 | 5 | 8 | 10 | 13 |
| 27 | 0.4540 | .4555 | .4571 | .4586 | .4602 | .4617 | .4633 | .4648 | .4664 | .4679 | 3 | 5 | 8 | 10 | 13 |
| 28 | 0.4695 | .4710 | .4726 | .4741 | .4756 | .4772 | .4787 | .4802 | .4818 | .4833 | 3 | 5 | 8 | 10 | 13 |
| 29 | 0.4848 | .4863 | .4879 | .4894 | .4909 | .4924 | .4939 | .4955 | .4970 | .4985 | 3 | 5 | 8 | 10 | 13 |
| 30 | 0.5000 | .5015 | .5030 | .5045 | .5060 | .5075 | .5090 | .5105 | .5120 | .5135 | 2 | 5 | 8 | 10 | 12 |
| 31 | 0.5150 | .5165 | .5180 | .5195 | .5210 | .5225 | .5240 | .5255 | .5270 | .5284 | 2 | 5 | 7 | 10 | 12 |
| 32 | 0.5299 | .5314 | .5329 | .5344 | .5358 | .5373 | .5388 | .5402 | .5417 | .5432 | 2 | 5 | 7 | 10 | 12 |
| 33 | 0.5446 | .5461 | .5476 | .5490 | .5505 | .5519 | .5534 | .5548 | .5563 | .5577 | 2 | 5 | 7 | 10 | 12 |
| 34 | 0.5592 | .5606 | .5621 | .5635 | .5650 | .5664 | .5678 | .5693 | .5707 | .5721 | 2 | 5 | 7 | 10 | 12 |
| 35 | 0.5736 | .5750 | .5764 | .5779 | .5793 | .5807 | .5821 | .5835 | .5850 | .5864 | 2 | 5 | 7 | 9 | 12 |
| 36 | 0.5878 | .5892 | .5906 | .5920 | .5934 | .5948 | .5962 | .5976 | .5990 | .6004 | 2 | 5 | 7 | 9 | 12 |
| 37 | 0.6018 | .6032 | .6046 | .6060 | .6074 | .6088 | .6101 | .6115 | .6129 | .6143 | 2 | 5 | 7 | 9 | 12 |
| 38 | 0.6157 | .6170 | .6184 | .6198 | .6211 | .6225 | .6239 | .6252 | .6266 | .6280 | 2 | 5 | 7 | 9 | 11 |
| 39 | 0.6293 | .6307 | .6320 | .6334 | .6347 | .6361 | .6374 | .6388 | .6401 | .6414 | 2 | 4 | 7 | 9 | 11 |
| 40 | 0.6428 | .6441 | .6455 | .6468 | .6481 | .6494 | .6508 | .6521 | .6534 | .6547 | 2 | 4 | 7 | 9 | 11 |
| 41 | 0.6561 | .6574 | .6587 | .6600 | .6613 | .6626 | .6639 | .6652 | .6665 | .6678 | 2 | 4 | 6 | 9 | 11 |
| 42 | 0.6691 | .6704 | .6717 | .6730 | .6743 | .6756 | .6769 | .6782 | .6794 | .6807 | 2 | 4 | 6 | 9 | 11 |
| 43 | 0.6820 | .6833 | .6845 | .6858 | .6871 | .6884 | .6896 | .6909 | .6921 | .6934 | 2 | 4 | 6 | 8 | 11 |
| 44 | 0.6947 | .6959 | .6972 | .6984 | .6997 | .7009 | .7022 | .7034 | .7046 | .7059 | 2 | 4 | 6 | 8 | 10 |
| | 0′ | 6′ | 12′ | 18′ | 24′ | 30′ | 36′ | 42′ | 48′ | 54′ | 1′ | 2′ | 3′ | 4′ | 5′ |

# NATURAL SINES

Proportional Parts

| | 0′ | 6′ 0.1 | 12′ 0.2 | 18′ 0.3 | 24′ 0.4 | 30′ 0.5 | 36′ 0.6 | 42′ 0.7 | 48′ 0.8 | 54′ 0.9 | 1′ | 2′ | 3′ | 4′ | 5′ |
|---|---|---|---|---|---|---|---|---|---|---|---|---|---|---|---|
| 45° | 0.7071 | .7083 | .7096 | .7108 | .7120 | .7133 | .7145 | .7157 | .7169 | .7181 | 2 | 4 | 6 | 8 | 10 |
| 46 | 0.7193 | .7206 | .7218 | .7230 | .7242 | .7254 | .7266 | .7278 | .7290 | .7302 | 2 | 4 | 6 | 8 | 10 |
| 47 | 0.7314 | .7325 | .7337 | .7349 | .7361 | .7373 | .7385 | .7396 | .7408 | .7420 | 2 | 4 | 6 | 8 | 10 |
| 48 | 0.7431 | .7443 | .7455 | .7466 | .7478 | .7490 | .7501 | .7513 | .7524 | .7536 | 2 | 4 | 6 | 8 | 10 |
| 49 | 0.7547 | .7559 | .7570 | .7581 | .7593 | .7604 | .7615 | .7627 | .7638 | .7649 | 2 | 4 | 6 | 8 | 9 |
| 50 | 0.7660 | .7672 | .7683 | .7694 | .7705 | .7716 | .7727 | .7738 | .7749 | .7760 | 2 | 4 | 6 | 7 | 9 |
| 51 | 0.7771 | .7782 | .7793 | .7804 | .7815 | .7826 | .7837 | .7848 | .7859 | .7869 | 2 | 4 | 5 | 7 | 9 |
| 52 | 0.7880 | .7891 | .7902 | .7912 | .7923 | .7934 | .7944 | .7955 | .7965 | .7976 | 2 | 4 | 5 | 7 | 9 |
| 53 | 0.7986 | .7997 | .8007 | .8018 | .8028 | .8039 | .8049 | .8059 | .8070 | .8080 | 2 | 3 | 5 | 7 | 9 |
| 54 | 0.8090 | .8100 | .8111 | .8121 | .8131 | .8141 | .8151 | .8161 | .8171 | .8181 | 2 | 3 | 5 | 7 | 8 |
| 55 | 0.8192 | .8202 | .8211 | .8221 | .8231 | .8241 | .8251 | .8261 | .8271 | .8281 | 2 | 3 | 5 | 7 | 8 |
| 56 | 0.8290 | .8300 | .8310 | .8320 | .8329 | .8339 | .8348 | .8358 | .8368 | .8377 | 2 | 3 | 5 | 6 | 8 |
| 57 | 0.8387 | .8396 | .8406 | .8415 | .8425 | .8434 | .8443 | .8453 | .8462 | .8471 | 2 | 3 | 5 | 6 | 8 |
| 58 | 0.8480 | .8490 | .8499 | .8508 | .8517 | .8526 | .8536 | .8545 | .8554 | .8563 | 2 | 3 | 5 | 6 | 8 |
| 59 | 0.8572 | .8581 | .8590 | .8599 | .8607 | .8616 | .8625 | .8634 | .8643 | .8652 | 1 | 3 | 4 | 6 | 7 |
| 60 | 0.8660 | .8669 | .8678 | .8686 | .8695 | .8704 | .8712 | .8721 | .8729 | .8738 | 1 | 3 | 4 | 6 | 7 |
| 61 | 0.8746 | .8755 | .8763 | .8771 | .8780 | .8788 | .8796 | .8805 | .8813 | .8821 | 1 | 3 | 4 | 6 | 7 |
| 62 | 0.8829 | .8838 | .8846 | .8854 | .8862 | .8870 | .8878 | .8886 | .8894 | .8902 | 1 | 3 | 4 | 5 | 7 |
| 63 | 0.8910 | .8918 | .8926 | .8934 | .8942 | .8949 | .8957 | .8965 | .8973 | .8980 | 1 | 3 | 4 | 5 | 6 |
| 64 | 0.8988 | .8996 | .9003 | .9011 | .9018 | .9026 | .9033 | .9041 | .9048 | .9056 | 1 | 2 | 4 | 5 | 6 |
| 65 | 0.9063 | .9070 | .9078 | .9085 | .9092 | .9100 | .9107 | .9114 | .9121 | .9128 | 1 | 2 | 4 | 5 | 6 |
| 66 | 0.9135 | .9143 | .9150 | .9157 | .9164 | .9171 | .9178 | .9184 | .9191 | .9198 | 1 | 2 | 3 | 5 | 6 |
| 67 | 0.9205 | .9212 | .9219 | .9225 | .9232 | .9239 | .9245 | .9252 | .9259 | .9265 | 1 | 2 | 3 | 4 | 6 |
| 68 | 0.9272 | .9278 | .9285 | .9291 | .9298 | .9304 | .9311 | .9317 | .9323 | .9330 | 1 | 2 | 3 | 4 | 5 |
| 69 | 0.9336 | .9342 | .9348 | .9354 | .9361 | .9367 | .9373 | .9379 | .9385 | .9391 | 1 | 2 | 3 | 4 | 5 |
| 70 | 0.9397 | .9403 | .9409 | .9415 | .9421 | .9426 | .9432 | .9438 | .9444 | .9449 | 1 | 2 | 3 | 4 | 5 |
| 71 | 0.9455 | .9461 | .9466 | .9472 | .9478 | .9483 | .9489 | .9494 | .9500 | .9505 | 1 | 2 | 3 | 4 | 5 |
| 72 | 0.9511 | .9516 | .9521 | .9527 | .9532 | .9537 | .9542 | .9548 | .9553 | .9558 | 1 | 2 | 3 | 3 | 4 |
| 73 | 0.9563 | .9568 | .9573 | .9578 | .9583 | .9588 | .9593 | .9598 | .9603 | .9608 | 1 | 2 | 2 | 3 | 4 |
| 74 | 0.9613 | .9617 | .9622 | .9627 | .9632 | .9636 | .9641 | .9646 | .9650 | .9655 | 1 | 2 | 2 | 3 | 4 |
| 75 | 0.9659 | .9664 | .9668 | .9673 | .9677 | .9681 | .9686 | .9690 | .9694 | .9699 | 1 | 1 | 2 | 3 | 4 |
| 76 | 0.9703 | .9707 | .9711 | .9715 | .9720 | .9724 | .9728 | .9732 | .9736 | .9740 | 1 | 1 | 2 | 3 | 3 |
| 77 | 0.9744 | .9748 | .9751 | .9755 | .9759 | .9763 | .9767 | .9770 | .9774 | .9778 | 1 | 1 | 2 | 2 | 3 |
| 78 | 0.9781 | .9785 | .9789 | .9792 | .9796 | .9799 | .9803 | .9806 | .9810 | .9813 | 1 | 1 | 2 | 2 | 3 |
| 79 | 0.9816 | .9820 | .9823 | .9826 | .9829 | .9833 | .9836 | .9839 | .9842 | .9845 | 1 | 1 | 2 | 2 | 3 |
| 80 | 0.9848 | .9851 | .9854 | .9857 | .9860 | .9863 | .9866 | .9869 | .9871 | .9874 | 0 | 1 | 1 | 2 | 2 |
| 81 | 0.9877 | .9880 | .9882 | .9885 | .9888 | .9890 | .9893 | .9895 | .9898 | .9900 | 0 | 1 | 1 | 2 | 2 |
| 82 | 0.9903 | .9905 | .9907 | .9910 | .9912 | .9914 | .9917 | .9919 | .9921 | .9923 | 0 | 1 | 1 | 1 | 2 |
| 83 | 0.9925 | .9928 | .9930 | .9932 | .9934 | .9936 | .9938 | .9940 | .9942 | .9943 | 0 | 1 | 1 | 1 | 2 |
| 84 | 0.9945 | .9947 | .9949 | .9951 | .9952 | .9954 | .9956 | .9957 | .9959 | .9960 | 0 | 1 | 1 | 1 | 1 |
| 85 | 0.9962 | .9963 | .9965 | .9966 | .9968 | .9969 | .9971 | .9972 | .9973 | .9974 | 0 | 0 | 1 | 1 | 1 |
| 86 | 0.9976 | .9977 | .9978 | .9979 | .9980 | .9981 | .9982 | .9983 | .9984 | .9985 | 0 | 0 | 0 | 1 | 1 |
| 87 | 0.9986 | .9987 | .9988 | .9989 | .9990 | .9990 | .9991 | .9992 | .9993 | .9993 | 0 | 0 | 0 | 1 | 1 |
| 88 | 0.9994 | .9995 | .9995 | .9996 | .9996 | .9997 | .9997 | .9997 | .9998 | .9998 | 0 | 0 | 0 | 0 | 0 |
| 89 | 0.9998 | .9999 | .9999 | .9999 | 0.9999 | 1.0000 | .0000 | .0000 | .0000 | .0000 | 0 | 0 | 0 | 0 | 0 |
| | 0′ | 6′ | 12′ | 18′ | 24′ | 30′ | 36′ | 42′ | 48′ | 54′ | 1′ | 2′ | 3′ | 4′ | 5′ |

# NATURAL COSINES

Proportional Parts Subtract

| | 0′ | 6′ 0.1 | 12′ 0.2 | 18′ 0.3 | 24′ 0.4 | 30′ 0.5 | 36′ 0.6 | 42′ 0.7 | 48′ 0.8 | 54′ 0.9 | 1′ | 2′ | 3′ | 4′ | 5′ |
|---|---|---|---|---|---|---|---|---|---|---|---|---|---|---|---|
| 0° | 1.0000 | .0000 | .0000 | .0000 | .0000 | 1.0000 | 0.9999 | .9999 | .9999 | .9999 | 0 | 0 | 0 | 0 | 0 |
| 1 | 0.9998 | .9998 | .9998 | .9997 | .9997 | .9997 | .9996 | .9996 | .9995 | .9995 | 0 | 0 | 0 | 0 | 0 |
| 2 | 0.9994 | .9993 | .9993 | .9992 | .9991 | .9990 | .9990 | .9989 | .9988 | .9987 | 0 | 0 | 0 | 0 | 1 |
| 3 | 0.9986 | .9985 | .9984 | .9983 | .9982 | .9981 | .9980 | .9979 | .9978 | .9977 | 0 | 0 | 0 | 1 | 1 |
| 4 | 0.9976 | .9974 | .9973 | .9972 | .9971 | .9969 | .9968 | .9966 | .9965 | .9963 | 0 | 0 | 1 | 1 | 1 |
| 5 | 0.9962 | .9960 | .9959 | .9957 | .9956 | .9954 | .9952 | .9951 | .9949 | .9947 | 0 | 1 | 1 | 1 | 1 |
| 6 | 0.9945 | .9943 | .9942 | .9940 | .9938 | .9936 | .9934 | .9932 | .9930 | .9928 | 0 | 1 | 1 | 1 | 2 |
| 7 | 0.9925 | .9923 | .9921 | .9919 | .9917 | .9914 | .9912 | .9910 | .9907 | .9905 | 0 | 1 | 1 | 1 | 2 |
| 8 | 0.9903 | .9900 | .9898 | .9895 | .9893 | .9890 | .9888 | .9885 | .9882 | .9880 | 0 | 1 | 1 | 2 | 2 |
| 9 | 0.9877 | .9874 | .9871 | .9869 | .9866 | .9863 | .9860 | .9857 | .9854 | .9851 | 0 | 1 | 1 | 2 | 2 |
| 10 | 0.9848 | .9845 | .9842 | .9839 | .9836 | .9833 | .9829 | .9826 | .9823 | .9820 | 1 | 1 | 2 | 2 | 3 |
| 11 | 0.9816 | .9813 | .9810 | .9806 | .9803 | .9799 | .9796 | .9792 | .9789 | .9785 | 1 | 1 | 2 | 2 | 3 |
| 12 | 0.9781 | .9778 | .9774 | .9770 | .9767 | .9763 | .9759 | .9755 | .9751 | .9748 | 1 | 1 | 2 | 2 | 3 |
| 13 | 0.9744 | .9740 | .9736 | .9732 | .9728 | .9724 | .9720 | .9715 | .9711 | .9707 | 1 | 1 | 2 | 3 | 3 |
| 14 | 0.9703 | .9699 | .9694 | .9690 | .9686 | .9681 | .9677 | .9673 | .9668 | .9664 | 1 | 1 | 2 | 3 | 4 |
| 15 | 0.9659 | .9655 | .9650 | .9646 | .9641 | .9636 | .9632 | .9627 | .9622 | .9617 | 1 | 2 | 2 | 3 | 4 |
| 16 | 0.9613 | .9608 | .9603 | .9598 | .9593 | .9588 | .9583 | .9578 | .9573 | .9568 | 1 | 2 | 2 | 3 | 4 |
| 17 | 0.9563 | .9558 | .9553 | .9548 | .9542 | .9537 | .9532 | .9527 | .9521 | .9516 | 1 | 2 | 3 | 3 | 4 |
| 18 | 0.9511 | .9505 | .9500 | .9494 | .9489 | .9483 | .9478 | .9472 | .9466 | .9461 | 1 | 2 | 3 | 4 | 5 |
| 19 | 0.9455 | .9449 | .9444 | .9438 | .9432 | .9426 | .9421 | .9415 | .9409 | .9403 | 1 | 2 | 3 | 4 | 5 |
| 20 | 0.9397 | .9391 | .9385 | .9379 | .9373 | .9367 | .9361 | .9354 | .9348 | .9342 | 1 | 2 | 3 | 4 | 5 |
| 21 | 0.9336 | .9330 | .9323 | .9317 | .9311 | .9304 | .9298 | .9291 | .9285 | .9278 | 1 | 2 | 3 | 4 | 5 |
| 22 | 0.9272 | .9265 | .9259 | .9252 | .9245 | .9239 | .9232 | .9225 | .9219 | .9212 | 1 | 2 | 3 | 4 | 6 |
| 23 | 0.9205 | .9198 | .9191 | .9184 | .9178 | .9171 | .9164 | .9157 | .9150 | .9143 | 1 | 2 | 3 | 5 | 6 |
| 24 | 0.9135 | .9128 | .9121 | .9114 | .9107 | .9100 | .9092 | .9085 | .9078 | .9070 | 1 | 2 | 4 | 5 | 6 |
| 25 | 0.9063 | .9056 | .9048 | .9041 | .9033 | .9026 | .9018 | .9011 | .9003 | .8996 | 1 | 2 | 4 | 5 | 6 |
| 26 | 0.8988 | .8980 | .8973 | .8965 | .8957 | .8949 | .8942 | .8934 | .8926 | .8918 | 1 | 3 | 4 | 5 | 6 |
| 27 | 0.8910 | .8902 | .8894 | .8886 | .8878 | .8870 | .8862 | .8854 | .8846 | .8838 | 1 | 3 | 4 | 5 | 7 |
| 28 | 0.8829 | .8821 | .8813 | .8805 | .8796 | .8788 | .8780 | .8771 | .8763 | .8755 | 1 | 3 | 4 | 6 | 7 |
| 29 | 0.8746 | .8738 | .8729 | .8721 | .8712 | .8704 | .8695 | .8686 | .8678 | .8669 | 1 | 3 | 4 | 6 | 7 |
| 30 | 0.8660 | .8652 | .8643 | .8634 | .8625 | .8616 | .8607 | .8599 | .8590 | .8581 | 1 | 3 | 4 | 6 | 7 |
| 31 | 0.8572 | .8563 | .8554 | .8545 | .8536 | .8526 | .8517 | .8508 | .8499 | .8490 | 2 | 3 | 5 | 6 | 8 |
| 32 | 0.8480 | .8471 | .8462 | .8453 | .8443 | .8434 | .8425 | .8415 | .8406 | .8396 | 2 | 3 | 5 | 6 | 8 |
| 33 | 0.8387 | .8377 | .8368 | .8358 | .8348 | .8339 | .8329 | .8320 | .8310 | .8300 | 2 | 3 | 5 | 6 | 8 |
| 34 | 0.8290 | .8281 | .8271 | .8261 | .8251 | .8241 | .8231 | .8221 | .8211 | .8202 | 2 | 3 | 5 | 7 | 8 |
| 35 | 0.8192 | .8181 | .8171 | .8161 | .8151 | .8141 | .8131 | .8121 | .8111 | .8100 | 2 | 3 | 5 | 7 | 8 |
| 36 | 0.8090 | .8080 | .8070 | .8059 | .8049 | .8039 | .8028 | .8018 | .8007 | .7997 | 2 | 3 | 5 | 7 | 9 |
| 37 | 0.7986 | .7976 | .7965 | .7955 | .7944 | .7934 | .7923 | .7912 | .7902 | .7891 | 2 | 4 | 5 | 7 | 9 |
| 38 | 0.7880 | .7869 | .7859 | .7848 | .7837 | .7826 | .7815 | .7804 | .7793 | .7782 | 2 | 4 | 5 | 7 | 9 |
| 39 | 0.7771 | .7760 | .7749 | .7738 | .7727 | .7716 | .7705 | .7694 | .7683 | .7672 | 2 | 4 | 6 | 7 | 9 |
| 40 | 0.7660 | .7649 | .7638 | .7627 | .7615 | .7604 | .7593 | .7581 | .7570 | .7559 | 2 | 4 | 6 | 8 | 9 |
| 41 | 0.7547 | .7536 | .7524 | .7513 | .7501 | .7490 | .7478 | .7466 | .7455 | .7443 | 2 | 4 | 6 | 8 | 10 |
| 42 | 0.7431 | .7420 | .7408 | .7396 | .7385 | .7373 | .7361 | .7349 | .7337 | .7325 | 2 | 4 | 6 | 8 | 10 |
| 43 | 0.7314 | .7302 | .7290 | .7278 | .7266 | .7254 | .7242 | .7230 | .7218 | .7206 | 2 | 4 | 6 | 8 | 10 |
| 44 | 0.7193 | .7181 | .7169 | .7157 | .7145 | .7133 | .7120 | .7108 | .7096 | .7083 | 2 | 4 | 6 | 8 | 10 |
| | 0′ | 6′ | 12′ | 18′ | 24′ | 30′ | 36′ | 42′ | 48′ | 54′ | 1′ | 2′ | 3′ | 4′ | 5′ |

# NATURAL COSINES

Proportional Parts Subtract

| | 0′ | 6′ 0.1 | 12′ 0.2 | 18′ 0.3 | 24′ 0.4 | 30′ 0.5 | 36′ 0.6 | 42′ 0.7 | 48′ 0.8 | 54′ 0.9 | 1′ | 2′ | 3′ | 4′ | 5′ |
|---|---|---|---|---|---|---|---|---|---|---|---|---|---|---|---|
| 45° | 0.7071 | .7059 | .7046 | .7034 | .7022 | .7009 | .6997 | .6984 | .6972 | .6959 | 2 | 4 | 6 | 8 | 10 |
| 46 | 0.6947 | .6934 | .6921 | .6909 | .6896 | .6884 | .6871 | .6858 | .6845 | .6833 | 2 | 4 | 6 | 8 | 11 |
| 47 | 0.6820 | .6807 | .6794 | .6782 | .6769 | .6756 | .6743 | .6730 | .6717 | .6704 | 2 | 4 | 6 | 9 | 11 |
| 48 | 0.6691 | .6678 | .6665 | .6652 | .6639 | .6626 | .6613 | .6600 | .6587 | .6574 | 2 | 4 | 6 | 9 | 11 |
| 49 | 0.6561 | .6547 | .6534 | .6521 | .6508 | .6494 | .6481 | .6468 | .6455 | .6441 | 2 | 4 | 7 | 9 | 11 |
| 50 | 0.6428 | .6414 | .6401 | .6388 | .6374 | .6361 | .6347 | .6334 | .6320 | .6307 | 2 | 4 | 7 | 9 | 11 |
| 51 | 0.6293 | .6280 | .6266 | .6252 | .6239 | .6225 | .6211 | .6198 | .6184 | .6170 | 2 | 5 | 7 | 9 | 11 |
| 52 | 0.6157 | .6143 | .6129 | .6115 | .6101 | .6088 | .6074 | .6060 | .6046 | .6032 | 2 | 5 | 7 | 9 | 12 |
| 53 | 0.6018 | .6004 | .5990 | .5976 | .5962 | .5948 | .5934 | .5920 | .5906 | .5892 | 2 | 5 | 7 | 9 | 12 |
| 54 | 0.5878 | .5864 | .5850 | .5835 | .5821 | .5807 | .5793 | .5779 | .5764 | .5750 | 2 | 5 | 7 | 9 | 12 |
| 55 | 0.5736 | .5721 | .5707 | .5693 | .5678 | .5664 | .5650 | .5635 | .5621 | .5606 | 2 | 5 | 7 | 10 | 12 |
| 56 | 0.5592 | .5577 | .5563 | .5548 | .5534 | .5519 | .5505 | .5490 | .5476 | .5461 | 2 | 5 | 7 | 10 | 12 |
| 57 | 0.5446 | .5432 | .5417 | .5402 | .5388 | .5373 | .5358 | .5344 | .5329 | .5314 | 2 | 5 | 7 | 10 | 12 |
| 58 | 0.5299 | .5284 | .5270 | .5255 | .5240 | .5225 | .5210 | .5195 | .5180 | .5165 | 2 | 5 | 7 | 10 | 12 |
| 59 | 0.5150 | .5135 | .5120 | .5105 | .5090 | .5075 | .5060 | .5045 | .5030 | .5015 | 2 | 5 | 8 | 10 | 12 |
| 60 | 0.5000 | .4985 | .4970 | .4955 | .4939 | .4924 | .4909 | .4894 | .4879 | .4863 | 3 | 5 | 8 | 10 | 13 |
| 61 | 0.4848 | .4833 | .4818 | .4802 | .4787 | .4772 | .4756 | .4741 | .4726 | .4710 | 3 | 5 | 8 | 10 | 13 |
| 62 | 0.4695 | .4679 | .4664 | .4648 | .4633 | .4617 | .4602 | .4586 | .4571 | .4555 | 3 | 5 | 8 | 10 | 13 |
| 63 | 0.4450 | .4524 | .4509 | .4493 | .4478 | .4462 | .4446 | .4431 | .4415 | .4399 | 3 | 5 | 8 | 10 | 13 |
| 64 | 0.4384 | .4368 | .4352 | .4337 | .4321 | .4305 | .4289 | .4274 | .4258 | .4242 | 3 | 5 | 8 | 11 | 13 |
| 65 | 0.4226 | .4210 | .4195 | .4179 | .4163 | .4147 | .4131 | .4115 | .4099 | .4083 | 3 | 5 | 8 | 11 | 13 |
| 66 | 0.4067 | .4051 | .4035 | .4019 | .4003 | .3987 | .3971 | .3955 | .3939 | .3923 | 3 | 5 | 8 | 11 | 13 |
| 67 | 0.3907 | .3891 | .3875 | .3859 | .3843 | .3827 | .3811 | .3795 | .3778 | .3762 | 3 | 5 | 8 | 11 | 13 |
| 68 | 0.3746 | .3730 | .3714 | .3697 | .3681 | .3665 | .3649 | .3633 | .3616 | .3600 | 3 | 5 | 8 | 11 | 14 |
| 69 | 0.3584 | .3567 | .3551 | .3535 | .3518 | .3502 | .3486 | .3469 | .3453 | .3437 | 3 | 5 | 8 | 11 | 14 |
| 70 | 0.3420 | .3404 | .3387 | .3371 | .3355 | .3338 | .3322 | .3305 | .3289 | .3272 | 3 | 5 | 8 | 11 | 14 |
| 71 | 0.3256 | .3239 | .3223 | .3206 | .3190 | .3173 | .3156 | .3140 | .3123 | .3107 | 3 | 6 | 8 | 11 | 14 |
| 72 | 0.3090 | .3074 | .3057 | .3040 | .3024 | .3007 | .3990 | .3974 | .3957 | .3940 | 3 | 6 | 8 | 11 | 14 |
| 73 | 0.2924 | .2907 | .2890 | .2874 | .2857 | .2840 | .2823 | .2807 | .2790 | .2773 | 3 | 6 | 8 | 11 | 14 |
| 74 | 0.2756 | .2740 | .2723 | .2706 | .2689 | .2672 | .2656 | .2639 | .2622 | .2605 | 3 | 6 | 8 | 11 | 14 |
| 75 | 0.2588 | .2571 | .2554 | .2538 | .2521 | .2504 | .2487 | .2470 | .2453 | .2436 | 3 | 6 | 8 | 11 | 14 |
| 76 | 0.2419 | .2402 | .2385 | .2368 | .2351 | .2334 | .2317 | .2300 | .2284 | .2267 | 3 | 6 | 8 | 11 | 14 |
| 77 | 0.2250 | .2232 | .2215 | .2198 | .2181 | .2164 | .2147 | .2130 | .2113 | .2096 | 3 | 6 | 9 | 11 | 14 |
| 78 | 0.2079 | .2062 | .2045 | .2028 | .2011 | .1994 | .1977 | .1959 | .1942 | .1925 | 3 | 6 | 9 | 11 | 14 |
| 79 | 0.1908 | .1891 | .1874 | .1857 | .1840 | .1822 | .1805 | .1788 | .1771 | .1754 | 3 | 6 | 9 | 11 | 14 |
| 80 | 0.1736 | .1719 | .1702 | .1685 | .1668 | .1650 | .1633 | .1616 | .1599 | .1582 | 3 | 6 | 9 | 11 | 14 |
| 81 | 0.1564 | .1547 | .1530 | .1513 | .1495 | .1478 | .1461 | .1444 | .1426 | .1409 | 3 | 6 | 9 | 11 | 14 |
| 82 | 0.1392 | .1374 | .1357 | .1340 | .1323 | .1305 | .1288 | .1271 | .1253 | .1236 | 3 | 6 | 9 | 12 | 14 |
| 83 | 0.1219 | .1201 | .1184 | .1167 | .1149 | .1132 | .1115 | .1097 | .1080 | .1063 | 3 | 6 | 9 | 12 | 14 |
| 84 | 0.1045 | .1028 | .1011 | .0993 | .0976 | .0958 | .0941 | .0924 | .0906 | .0889 | 3 | 6 | 9 | 12 | 14 |
| 85 | 0.0872 | .0854 | .0837 | .0819 | .0802 | .0085 | .0767 | .0750 | .0732 | .0715 | 3 | 6 | 9 | 12 | 14 |
| 86 | 0.0698 | .0680 | .0663 | .0645 | .0628 | .0610 | .0593 | .0576 | .0558 | .0541 | 3 | 6 | 9 | 12 | 15 |
| 87 | 0.0523 | .0506 | .0489 | .0471 | .0454 | .0436 | .0419 | .0401 | .0384 | .0366 | 3 | 6 | 9 | 12 | 15 |
| 88 | 0.0349 | .0332 | .0314 | .0297 | .0279 | .0262 | .0244 | .0227 | .0209 | .0192 | 3 | 6 | 9 | 12 | 15 |
| 89 | 0.0175 | .0157 | .0140 | .0122 | .0105 | .0087 | .0070 | .0052 | .0035 | .0017 | 3 | 6 | 9 | 12 | 15 |
| | 0′ | 6′ | 12′ | 18′ | 24′ | 30′ | 36′ | 42′ | 48′ | 54′ | 1′ | 2′ | 3′ | 4′ | 5′ |

# NATURAL COSECANTS

Proportional Parts Subtract

| | 0′ | 6′ 0.1 | 12′ 0.2 | 18′ 0.3 | 24′ 0.4 | 30′ 0.5 | 36′ 0.6 | 42′ 0.7 | 48′ 0.8 | 54′ 0.9 | 1′ | 2′ | 3′ | 4′ | 5′ |
|---|---|---|---|---|---|---|---|---|---|---|---|---|---|---|---|
| 0° | ∞ | 573.0 | 286.5 | 191.0 | 143.2 | 114.6 | 95.49 | 81.85 | 71.62 | 63.66 | p.p. cease to be sufficiently accurate | | | | |
| 1 | 57.30 | 52.09 | 47.75 | 44.08 | 40.93 | 38.20 | 35.81 | 33.71 | 31.84 | 30.16 | | | | | |
| 2 | 28.65 | 27.29 | 26.05 | 24.92 | 23.88 | 22.93 | 22.04 | 21.23 | 20.47 | 19.77 | | | | | |
| 3 | 19.11 | 18.49 | 17.91 | 17.37 | 16.86 | 16.38 | 15.93 | 15.50 | 15.09 | 14.70 | | | | | |
| 4 | 14.34 | 13.99 | 13.65 | 13.34 | 13.03 | 12.75 | 12.47 | 12.20 | 11.95 | 11.71 | | | | | |
| 5 | 11.474 | .249 | 11.034 | 10.826 | .626 | .433 | .248 | 10.068 | 9.895 | .728 | | | | | |
| 6 | 9.567 | .411 | .259 | 9.113 | 8.971 | .834 | .700 | .571 | .446 | .324 | | | | | |
| 7 | 8.206 | 8.091 | 7.979 | .870 | .764 | .661 | .561 | .463 | .368 | .276 | | | | | |
| 8 | 7.185 | .097 | 7.011 | 6.927 | .845 | .765 | .687 | .611 | .537 | .464 | | | | | |
| 9 | 6.392 | .323 | .255 | .188 | .123 | 6.059 | 5.996 | .935 | .875 | .816 | | | | | |
| 10 | 5.759 | .702 | .647 | .593 | .540 | .487 | .436 | .386 | .337 | .288 | 9 | 17 | 26 | 35 | 43 |
| 11 | 5.241 | .194 | .148 | .103 | .059 | 5.016 | 4.973 | .931 | .890 | .850 | 7 | 14 | 22 | 29 | 36 |
| 12 | 4.810 | .771 | .732 | .694 | .657 | .620 | .584 | .549 | .514 | .479 | 6 | 12 | 18 | 24 | 30 |
| 13 | 4.445 | .412 | .379 | .347 | .315 | .284 | .253 | .222 | .192 | .163 | 5 | 10 | 16 | 21 | 26 |
| 14 | 4.134 | .105 | .077 | .049 | 4.021 | 3.994 | .967 | .941 | .915 | .889 | 4 | 9 | 14 | 18 | 22 |
| 15 | 3.864 | .839 | .814 | .790 | .766 | .742 | .719 | .695 | .673 | .650 | 4 | 8 | 12 | 16 | 20 |
| 16 | 3.628 | .606 | .584 | .563 | .542 | .521 | .500 | .480 | .460 | .440 | 3 | 7 | 10 | 14 | 17 |
| 17 | 3.420 | .401 | .382 | .363 | .344 | .326 | .307 | .289 | .271 | .254 | 3 | 6 | 9 | 12 | 15 |
| 18 | 3.236 | .219 | .202 | .185 | .168 | .152 | .135 | .119 | .103 | .087 | 3 | 5 | 8 | 11 | 14 |
| 19 | 3.072 | .056 | .041 | .026 | 3.011 | 2.996 | .981 | .967 | .952 | .938 | 2 | 5 | 7 | 10 | 12 |
| 20 | 2.924 | .910 | .896 | .882 | .869 | .855 | .842 | .829 | .816 | .803 | 2 | 4 | 7 | 9 | 11 |
| 21 | 2.790 | .778 | .765 | .753 | .741 | .729 | .716 | .705 | .693 | .681 | 2 | 4 | 6 | 8 | 10 |
| 22 | 2.669 | .658 | .647 | .635 | .624 | .613 | .602 | .591 | .581 | .570 | 2 | 4 | 6 | 7 | 9 |
| 23 | 2.559 | .549 | .538 | .528 | .518 | .508 | .498 | .488 | .478 | .468 | 2 | 3 | 5 | 7 | 8 |
| 24 | 2.459 | .449 | .439 | .430 | .421 | .411 | .402 | .393 | .384 | .375 | 2 | 3 | 5 | 6 | 8 |
| 25 | 2.366 | .357 | .349 | .340 | .331 | .323 | .314 | .306 | .298 | .289 | 1 | 3 | 4 | 6 | 7 |
| 26 | 2.281 | .273 | .265 | .257 | .249 | .241 | .233 | .226 | .218 | .210 | 1 | 3 | 4 | 5 | 7 |
| 27 | 2.203 | .195 | .188 | .180 | .173 | .166 | .158 | .151 | .144 | .137 | 1 | 2 | 4 | 5 | 6 |
| 28 | 2.130 | .123 | .116 | .109 | .103 | .096 | .089 | .082 | .076 | .069 | 1 | 2 | 3 | 4 | 6 |
| 29 | 2.063 | .056 | .050 | .043 | .037 | .031 | .025 | .018 | .012 | .006 | 1 | 2 | 3 | 4 | 5 |
| 30 | 2.000 | 1.9940 | .9880 | .9821 | .9762 | .9703 | .9645 | .9587 | .9530 | .9473 | 10 | 19 | 29 | 39 | 49 |
| 31 | 1.9416 | .9360 | .9304 | .9249 | .9194 | .9139 | .9084 | .9031 | .8977 | .8924 | 9 | 18 | 27 | 36 | 45 |
| 32 | 1.8871 | .8818 | .8766 | .8714 | .8663 | .8612 | .8561 | .8510 | .8460 | .8410 | 8 | 17 | 26 | 34 | 42 |
| 33 | 1.8361 | .8312 | .8263 | .8214 | .8166 | .8118 | .8070 | .8023 | .7976 | .7929 | 8 | 16 | 24 | 32 | 40 |
| 34 | 1.7883 | .7837 | .7791 | .7745 | .7700 | .7655 | .7610 | .7566 | .7522 | .7478 | 7 | 15 | 22 | 30 | 37 |
| 35 | 1.7434 | .7391 | .7348 | .7305 | .7263 | .7221 | .7179 | .7137 | .7095 | .7054 | 7 | 14 | 21 | 28 | 35 |
| 36 | 1.7013 | .6972 | .6932 | .6892 | .6852 | .6812 | .6772 | .6733 | .6694 | .6655 | 7 | 13 | 20 | 26 | 33 |
| 37 | 1.6616 | .6578 | .6540 | .6502 | .6464 | .6427 | .6390 | .6353 | .6316 | .6279 | 6 | 12 | 19 | 25 | 31 |
| 38 | 1.6243 | .6207 | .6171 | .6135 | .6099 | .6064 | .6029 | .5994 | .5959 | .5925 | 6 | 12 | 18 | 24 | 29 |
| 39 | 1.5890 | .5856 | .5822 | .5788 | .5755 | .5721 | .5688 | .5655 | .5622 | .5590 | 6 | 11 | 17 | 22 | 28 |
| 40 | 1.5557 | .5525 | .5493 | .5461 | .5429 | .5398 | .5366 | .5335 | .5304 | .5273 | 5 | 10 | 16 | 21 | 26 |
| 41 | 1.5243 | .5212 | .5182 | .5151 | .5121 | .5092 | .5062 | .5032 | .5003 | .4974 | 5 | 10 | 15 | 20 | 25 |
| 42 | 1.4945 | .4916 | .4887 | .4859 | .4830 | .4802 | .4774 | .4746 | .4718 | .4690 | 5 | 9 | 14 | 19 | 24 |
| 43 | 1.4663 | .4635 | .4608 | .4581 | .4554 | .4527 | .4501 | .4474 | .4448 | .4422 | 4 | 9 | 13 | 18 | 22 |
| 44 | 1.4396 | .4370 | .4344 | .4318 | .4293 | .4267 | .4242 | .4217 | .4192 | .4167 | 4 | 8 | 13 | 17 | 21 |
| | 0′ | 6′ | 12′ | 18′ | 24′ | 30′ | 36′ | 42′ | 48′ | 54′ | 1′ | 2′ | 3′ | 4′ | 5′ |

# NATURAL COSECANTS

Proportional Parts Subtract

| | 0′ | 6′ 0.1 | 12′ 0.2 | 18′ 0.3 | 24′ 0.4 | 30′ 0.5 | 36′ 0.6 | 42′ 0.7 | 48′ 0.8 | 54′ 0.9 | 1′ | 2′ | 3′ | 4′ | 5′ |
|---|---|---|---|---|---|---|---|---|---|---|---|---|---|---|---|
| 45° | 1.4142 | .4118 | .4093 | .4069 | .4044 | .4020 | .3996 | .3972 | .3949 | .3925 | 4 | 8 | 12 | 16 | 20 |
| 46 | 1.3902 | .3878 | .3855 | .3832 | .3809 | .3786 | .3763 | .3741 | .3718 | .3696 | 4 | 8 | 11 | 15 | 19 |
| 47 | 1.3673 | .3651 | .3629 | .3607 | .3585 | .3563 | .3542 | .3520 | .3499 | .3478 | 4 | 7 | 11 | 14 | 18 |
| 48 | 1.3456 | .3435 | .3414 | .3393 | .3373 | .3352 | .3331 | .3311 | .3291 | .3270 | 3 | 7 | 10 | 14 | 17 |
| 49 | 1.3250 | .3230 | .3210 | .3190 | .3171 | .3151 | .3131 | .3112 | .3093 | .3073 | 3 | 7 | 10 | 13 | 16 |
| 50 | 1.3054 | .3035 | .3016 | .3997 | .3978 | .3960 | .3941 | .3923 | .3904 | .3886 | 3 | 6 | 9 | 12 | 15 |
| 51 | 1.2868 | .2849 | .2831 | .2813 | .2796 | .2778 | .2760 | .2742 | .2725 | .2708 | 3 | 6 | 9 | 12 | 15 |
| 52 | 1.2690 | .2673 | .2656 | .2639 | .2622 | .2605 | .2588 | .2571 | .2554 | .2538 | 3 | 6 | 8 | 11 | 14 |
| 53 | 1.2521 | .2505 | .2489 | .2472 | .2456 | .2440 | .2424 | .2408 | .2392 | .2376 | 3 | 5 | 8 | 11 | 13 |
| 54 | 1.2361 | .2345 | .2329 | .2314 | .2299 | .2283 | .2268 | .2253 | .2238 | .2223 | 3 | 5 | 8 | 10 | 13 |
| 55 | 1.2208 | .2193 | .2178 | .2163 | .2149 | .2134 | .2120 | .2105 | .2091 | .2076 | 2 | 5 | 7 | 10 | 12 |
| 56 | 1.2062 | .2048 | .2034 | .2020 | .2006 | .1992 | .1978 | .1964 | .1951 | .1937 | 2 | 5 | 7 | 9 | 12 |
| 57 | 1.1924 | .1910 | .1897 | .1883 | .1870 | .1857 | .1844 | .1831 | .1818 | .1805 | 2 | 4 | 7 | 9 | 11 |
| 58 | 1.1792 | .1779 | .1766 | .1753 | .1741 | .1728 | .1716 | .1703 | .1691 | .1679 | 2 | 4 | 6 | 8 | 10 |
| 59 | 1.1666 | .1654 | .1642 | .1630 | .1618 | .1606 | .1594 | .1582 | .1570 | .1559 | 2 | 4 | 6 | 8 | 10 |
| 60 | 1.1547 | .1535 | .1524 | .1512 | .1501 | .1490 | .1478 | .1467 | .1456 | .1445 | 2 | 4 | 6 | 8 | 9 |
| 61 | 1.1434 | .1423 | .1412 | .1401 | .1390 | .1379 | .1368 | .1357 | .1347 | .1336 | 2 | 4 | 5 | 7 | 9 |
| 62 | 1.1326 | .1315 | .1305 | .1294 | .1284 | .1274 | .1264 | .1253 | .1243 | .1233 | 2 | 3 | 5 | 7 | 9 |
| 63 | 1.1223 | .1213 | .1203 | .1194 | .1184 | .1174 | .1164 | .1155 | .1145 | .1136 | 2 | 3 | 5 | 6 | 8 |
| 64 | 1.1126 | .1117 | .1107 | .1098 | .1089 | .1079 | .1070 | .1061 | .1052 | .1043 | 2 | 3 | 5 | 6 | 8 |
| 65 | 1.0346 | .1025 | .1016 | .1007 | .0998 | .0989 | .0981 | .0972 | .0963 | .0955 | 1 | 3 | 4 | 6 | 7 |
| 66 | 1.0946 | .0938 | .0929 | .0921 | .0913 | .0904 | .0896 | .0888 | .0880 | .0872 | 1 | 3 | 4 | 5 | 7 |
| 67 | 1.0864 | .0856 | .0848 | .0840 | .0832 | .0824 | .0816 | .0808 | .0801 | .0793 | 1 | 3 | 4 | 5 | 7 |
| 68 | 1.0785 | .0778 | .0770 | .0763 | .0755 | .0748 | .0740 | .0733 | .0726 | .0719 | 1 | 2 | 4 | 5 | 6 |
| 69 | 1.0711 | .0704 | .0697 | .0690 | .0683 | .0676 | .0669 | .0662 | .0655 | .0649 | 1 | 2 | 3 | 5 | 6 |
| 70 | 1.0642 | .0635 | .0628 | .0622 | .0615 | .0608 | .0602 | .0595 | .0589 | .0583 | 1 | 2 | 3 | 4 | 5 |
| 71 | 1.0576 | .0570 | .0564 | .0557 | .0551 | .0545 | .0539 | .0533 | .0527 | .0521 | 1 | 2 | 3 | 4 | 5 |
| 72 | 1.0515 | .0509 | .0503 | .0497 | .0491 | .0485 | .0480 | .0474 | .0468 | .0463 | 1 | 2 | 3 | 4 | 5 |
| 73 | 1.0457 | .0451 | .0446 | .0440 | .0435 | .0429 | .0424 | .0419 | .0413 | .0408 | 1 | 2 | 3 | 4 | 5 |
| 74 | 1.0403 | .0398 | .0393 | .0388 | .0382 | .0377 | .0372 | .0367 | .0363 | .0358 | 1 | 2 | 3 | 3 | 4 |
| 75 | 1.0353 | .0348 | .0343 | .0338 | .0334 | .0329 | .0324 | .0320 | .0315 | .0311 | 1 | 2 | 2 | 3 | 4 |
| 76 | 1.0306 | .0302 | .0297 | .0293 | .0288 | .0284 | .0280 | .0276 | .0271 | .0267 | 1 | 1 | 2 | 3 | 4 |
| 77 | 1.0263 | .0259 | .0255 | .0251 | .0247 | .0243 | .0239 | .0235 | .0231 | .0227 | 1 | 1 | 2 | 3 | 3 |
| 78 | 1.0223 | .0220 | .0216 | .0212 | .0209 | .0205 | .0201 | .0198 | .0194 | .0191 | 1 | 1 | 2 | 2 | 3 |
| 79 | 1.0187 | .0184 | .0180 | .0177 | .0174 | .0170 | .0167 | .0164 | .0161 | .0157 | 1 | 1 | 2 | 2 | 3 |
| 80 | 1.0154 | .0151 | .0148 | .0145 | .0142 | .0139 | .0136 | .0133 | .0130 | .0127 | 0 | 1 | 1 | 2 | 2 |
| 81 | 1.0125 | .0122 | .0119 | .0116 | .0114 | .0111 | .0108 | .0106 | .0103 | .0101 | 0 | 1 | 1 | 2 | 2 |
| 82 | 1.0098 | .0096 | .0093 | .0091 | .0089 | .0086 | .0084 | .0082 | .0079 | .0077 | 0 | 1 | 1 | 2 | 2 |
| 83 | 1.0075 | .0073 | .0071 | .0069 | .0067 | .0065 | .0063 | .0061 | .0059 | .0057 | 0 | 1 | 1 | 1 | 2 |
| 84 | 1.0055 | .0053 | .0051 | .0050 | .0048 | .0046 | .0045 | .0043 | .0041 | .0040 | 0 | 1 | 1 | 1 | 1 |
| 85 | 1.0038 | .0037 | .0035 | .0034 | .0032 | .0031 | .0030 | .0028 | .0027 | .0026 | 0 | 0 | 1 | 1 | 1 |
| 86 | 1.0024 | .0023 | .0022 | .0021 | .0020 | .0019 | .0018 | .0017 | .0016 | .0015 | 0 | 0 | 0 | 1 | 1 |
| 87 | 1.0014 | .0013 | .0012 | .0011 | .0010 | .0010 | .0009 | .0008 | .0007 | .0007 | 0 | 0 | 0 | 1 | 1 |
| 88 | 1.0006 | .0005 | .0005 | .0004 | .0004 | .0003 | .0003 | .0003 | .0002 | .0002 | 0 | 0 | 0 | 0 | 0 |
| 89 | 1.0002 | .0001 | .0001 | .0001 | .0001 | .0000 | .0000 | .0000 | .0000 | .0000 | 0 | 0 | 0 | 0 | 0 |
| | 0′ | 6′ | 12′ | 18′ | 24′ | 30′ | 36′ | 42′ | 48′ | 54′ | 1′ | 2′ | 3′ | 4′ | 5′ |

# NATURAL SECANTS

Proportional Parts

| | 0′ | 6′ 0.1 | 12′ 0.2 | 18′ 0.3 | 24′ 0.4 | 30′ 0.5 | 36′ 0.6 | 42′ 0.7 | 48′ 0.8 | 54′ 0.9 | 1′ | 2′ | 3′ | 4′ | 5′ |
|---|---|---|---|---|---|---|---|---|---|---|---|---|---|---|---|
| 0° | 1.0000 | .0000 | .0000 | .0000 | .0000 | .0000 | .0001 | .0001 | .0001 | .0001 | 0 | 0 | 0 | 0 | 0 |
| 1 | 1.0002 | .0002 | .0002 | .0003 | .0003 | .0003 | .0004 | .0004 | .0005 | .0005 | 0 | 0 | 0 | 0 | 0 |
| 2 | 1.0006 | .0007 | .0007 | .0008 | .0009 | .0010 | .0010 | .0011 | .0012 | .0013 | 0 | 0 | 0 | 1 | 1 |
| 3 | 1.0014 | .0015 | .0016 | .0017 | .0018 | .0019 | .0020 | .0021 | .0022 | .0023 | 0 | 0 | 0 | 1 | 1 |
| 4 | 1.0024 | .0026 | .0027 | .0028 | .0030 | .0031 | .0032 | .0034 | .0035 | .0037 | 0 | 0 | 1 | 1 | 1 |
| 5 | 1.0038 | .0040 | .0041 | .0043 | .0045 | .0046 | .0048 | .0050 | .0051 | .0053 | 0 | 1 | 1 | 1 | 1 |
| 6 | 1.0055 | .0057 | .0059 | .0061 | .0063 | .0065 | .0067 | .0069 | .0071 | .0073 | 0 | 1 | 1 | 1 | 2 |
| 7 | 1.0075 | .0077 | .0079 | .0082 | .0084 | .0086 | .0089 | .0091 | .0093 | .0096 | 0 | 1 | 1 | 2 | 2 |
| 8 | 1.0098 | .0101 | .0103 | .0106 | .0108 | .0111 | .0114 | .0116 | .0119 | .0122 | 0 | 1 | 1 | 2 | 2 |
| 9 | 1.0125 | .0127 | .0130 | .0133 | .0136 | .0139 | .0142 | .0145 | .0148 | .0151 | 0 | 1 | 1 | 2 | 2 |
| 10 | 1.0154 | .0157 | .0161 | .0164 | .0167 | .0170 | .0174 | .0177 | .0180 | .0184 | 1 | 1 | 2 | 2 | 3 |
| 11 | 1.0187 | .0191 | .0194 | .0198 | .0201 | .0205 | .0209 | .0212 | .0216 | .0220 | 1 | 1 | 2 | 2 | 3 |
| 12 | 1.0223 | .0227 | .0231 | .0235 | .0239 | .0243 | .0247 | .0251 | .0255 | .0259 | 1 | 1 | 2 | 3 | 3 |
| 13 | 1.0263 | .0267 | .0271 | .0276 | .0280 | .0284 | .0288 | .0293 | .0297 | .0302 | 1 | 1 | 2 | 3 | 4 |
| 14 | 1.0306 | .0311 | .0315 | .0320 | .0324 | .0329 | .0334 | .0338 | .0343 | .0348 | 1 | 2 | 2 | 3 | 4 |
| 15 | 1.0353 | .0358 | .0363 | .0367 | .0372 | .0377 | .0382 | .0388 | .0393 | .0398 | 1 | 2 | 3 | 3 | 4 |
| 16 | 1.0403 | .0408 | .0413 | .0419 | .0424 | .0429 | .0435 | .0440 | .0446 | .0451 | 1 | 2 | 3 | 4 | 5 |
| 17 | 1.0457 | .0463 | .0468 | .0474 | .0480 | .0485 | .0491 | .0497 | .0503 | .0509 | 1 | 2 | 3 | 4 | 5 |
| 18 | 1.0515 | .0521 | .0527 | .0533 | .0539 | .0545 | .0551 | .0557 | .0564 | .0570 | 1 | 2 | 3 | 4 | 5 |
| 19 | 1.0576 | .0583 | .0589 | .0595 | .0602 | .0608 | .0615 | .0622 | .0628 | .0635 | 1 | 2 | 3 | 4 | 5 |
| 20 | 1.0642 | .0649 | .0655 | .0662 | .0669 | .0676 | .0683 | .0690 | .0697 | .0704 | 1 | 2 | 3 | 5 | 6 |
| 21 | 1.0711 | .0719 | .0726 | .0733 | .0740 | .0748 | .0755 | .0763 | .0770 | .0778 | 1 | 2 | 4 | 5 | 6 |
| 22 | 1.0785 | .0793 | .0801 | .0808 | .0816 | .0824 | .0832 | .0840 | .0848 | .0856 | 1 | 3 | 4 | 5 | 7 |
| 23 | 1.0864 | .0872 | .0880 | .0888 | .0896 | .0904 | .0913 | .0921 | .0929 | .0938 | 1 | 3 | 4 | 5 | 7 |
| 24 | 1.0946 | .0955 | .0963 | .0972 | .0981 | .0989 | .0998 | .1007 | .1016 | .1025 | 1 | 3 | 4 | 6 | 7 |
| 25 | 1.1034 | .1043 | .1052 | .1061 | .1070 | .1079 | .1089 | .1098 | .1107 | .1117 | 2 | 3 | 5 | 6 | 8 |
| 26 | 1.1126 | .1136 | .1145 | .1155 | .1164 | .1174 | .1184 | .1194 | .1203 | .1213 | 2 | 3 | 5 | 6 | 8 |
| 27 | 1.1223 | .1233 | .1243 | .1253 | .1264 | .1274 | .1284 | .1294 | .1305 | .1315 | 2 | 3 | 5 | 7 | 9 |
| 28 | 1.1326 | .1336 | .1347 | .1357 | .1368 | .1379 | .1390 | .1401 | .1412 | .1423 | 2 | 4 | 5 | 7 | 9 |
| 29 | 1.1434 | .1445 | .1456 | .1467 | .1478 | .1490 | .1501 | .1512 | .1524 | .1535 | 2 | 4 | 6 | 8 | 9 |
| 30 | 1.1547 | .1559 | .1570 | .1582 | .1594 | .1606 | .1618 | .1630 | .1642 | .1654 | 2 | 4 | 6 | 8 | 10 |
| 31 | 1.1666 | .1679 | .1691 | .1703 | .1716 | .1728 | .1741 | .1753 | .1766 | .1779 | 2 | 4 | 6 | 8 | 10 |
| 32 | 1.1792 | .1805 | .1818 | .1831 | .1844 | .1857 | .1870 | .1883 | .1897 | .1910 | 2 | 4 | 7 | 9 | 11 |
| 33 | 1.1924 | .1937 | .1951 | .1964 | .1978 | .1992 | .2006 | .2020 | .2034 | .2048 | 2 | 5 | 7 | 9 | 12 |
| 34 | 1.2062 | .2076 | .2091 | .2105 | .2120 | .2134 | .2149 | .2163 | .2178 | .2193 | 2 | 5 | 7 | 10 | 12 |
| 35 | 1.2208 | .2223 | .2238 | .2253 | .2268 | .2283 | .2299 | .2314 | .2329 | .2345 | 3 | 5 | 8 | 10 | 13 |
| 36 | 1.2361 | .2376 | .2392 | .2408 | .2424 | .2440 | .2456 | .2472 | .2489 | .2505 | 3 | 5 | 8 | 11 | 13 |
| 37 | 1.2521 | .2538 | .2554 | .2571 | .2588 | .2605 | .2622 | .2639 | .2656 | .2673 | 3 | 6 | 8 | 11 | 14 |
| 38 | 1.2690 | .2708 | .2725 | .2742 | .2760 | .2778 | .2796 | .2813 | .2831 | .2849 | 3 | 6 | 9 | 12 | 15 |
| 39 | 1.2868 | .2886 | .2904 | .2923 | .2941 | .2960 | .2978 | .2997 | .3016 | .3035 | 3 | 6 | 9 | 12 | 15 |
| 40 | 1.3054 | .3073 | .3093 | .3112 | .3131 | .3151 | .3171 | .3190 | .3210 | .3230 | 3 | 7 | 10 | 13 | 16 |
| 41 | 1.3250 | .3270 | .3291 | .3311 | .3331 | .3352 | .3373 | .3393 | .3414 | .3435 | 3 | 7 | 10 | 14 | 17 |
| 42 | 1.3456 | .3478 | .3499 | .3520 | .3542 | .3563 | .3585 | .3607 | .3629 | .3651 | 4 | 7 | 11 | 14 | 18 |
| 43 | 1.3673 | .3696 | .3718 | .3741 | .3763 | .3786 | .3809 | .3832 | .3855 | .3878 | 4 | 8 | 11 | 15 | 19 |
| 44 | 1.3902 | .3925 | .3949 | .3972 | .3996 | .4020 | .4044 | .4069 | .4093 | .4118 | 4 | 8 | 12 | 16 | 20 |
| | 0′ | 6′ | 12′ | 18′ | 24′ | 30′ | 36′ | 42′ | 48′ | 54′ | 1′ | 2′ | 3′ | 4′ | 5′ |

# NATURAL SECANTS

Proportional Parts

| | 0′ | 6′ 0.1 | 12′ 0.2 | 18′ 0.3 | 24′ 0.4 | 30′ 0.5 | 36′ 0.6 | 42′ 0.7 | 48′ 0.8 | 54′ 0.9 | 1′ | 2′ | 3′ | 4′ | 5′ |
|---|---|---|---|---|---|---|---|---|---|---|---|---|---|---|---|
| 45° | 1.4142 | .4167 | .4192 | .4217 | .4242 | .4267 | .4293 | .4318 | .4344 | .4370 | 4 | 8 | 13 | 17 | 21 |
| 46 | 1.4396 | .4422 | .4448 | .4474 | .4501 | .4527 | .4554 | .4581 | .4608 | .4635 | 4 | 9 | 13 | 18 | 22 |
| 47 | 1.4663 | .4690 | .4718 | .4746 | .4774 | .4802 | .4830 | .4859 | .4887 | .4916 | 5 | 9 | 14 | 19 | 24 |
| 48 | 1.4945 | .4974 | .5003 | .5032 | .5062 | .5092 | .5121 | .5151 | .5182 | .5212 | 5 | 10 | 15 | 20 | 25 |
| 49 | 1.5243 | .5273 | .5304 | .5335 | .5366 | .5398 | .5429 | .5461 | .5493 | .5525 | 5 | 10 | 16 | 21 | 26 |
| 50 | 1.5557 | .5590 | .5622 | .5655 | .5688 | .5721 | .5755 | .5788 | .5822 | .5856 | 6 | 11 | 17 | 22 | 28 |
| 51 | 1.5890 | .5925 | .5959 | .5994 | .6029 | .6064 | .6099 | .6135 | .6171 | .6207 | 6 | 12 | 18 | 24 | 29 |
| 52 | 1.6243 | .6279 | .6316 | .6353 | .6390 | .6427 | .6464 | .6502 | .6540 | .6578 | 6 | 12 | 19 | 25 | 31 |
| 53 | 1.6616 | .6655 | .6694 | .6733 | .6772 | .6812 | .6852 | .6892 | .6932 | .6972 | 7 | 13 | 20 | 26 | 33 |
| 54 | 1.7013 | .7054 | .7095 | .7137 | .7179 | .7221 | .7263 | .7305 | .7348 | .7391 | 7 | 14 | 21 | 28 | 35 |
| 55 | 1.7434 | .7478 | .7522 | .7566 | .7610 | .7655 | .7700 | .7745 | .7791 | .7837 | 7 | 15 | 22 | 30 | 37 |
| 56 | 1.7883 | .7929 | .7976 | .8023 | .8070 | .8118 | .8166 | .8214 | .8263 | .8312 | 8 | 16 | 24 | 32 | 40 |
| 57 | 1.8361 | .8410 | .8460 | .8510 | .8561 | .8612 | .8663 | .8714 | .8766 | .8818 | 8 | 17 | 26 | 34 | 42 |
| 58 | 1.8871 | .8924 | .8977 | .9031 | .9084 | .9139 | .9194 | .9249 | .9304 | .9360 | 9 | 18 | 27 | 36 | 45 |
| 59 | 1.9416 | .9473 | .9530 | .9587 | .9645 | .9703 | .9762 | .9821 | .9880 | .9940 | 10 | 19 | 29 | 39 | 49 |
| 60 | 2.0000 | .006 | .012 | .018 | .025 | .031 | .037 | .043 | .050 | .056 | 1 | 2 | 3 | 4 | 5 |
| 61 | 2.063 | .069 | .076 | .082 | .089 | .096 | .103 | .109 | .116 | .123 | 1 | 2 | 3 | 4 | 6 |
| 62 | 2.130 | .137 | .144 | .151 | .158 | .166 | .173 | .180 | .188 | .195 | 1 | 2 | 4 | 5 | 6 |
| 63 | 2.203 | .210 | .218 | .226 | .233 | .241 | .249 | .257 | .265 | .273 | 1 | 3 | 4 | 5 | 7 |
| 64 | 2.281 | .289 | .298 | .306 | .314 | .323 | .331 | .340 | .349 | .357 | 1 | 3 | 4 | 6 | 7 |
| 65 | 2.366 | .375 | .384 | .393 | .402 | .411 | .421 | .430 | .439 | .449 | 2 | 3 | 5 | 6 | 8 |
| 66 | 2.459 | .468 | .478 | .488 | .498 | .508 | .518 | .528 | .538 | .549 | 2 | 3 | 5 | 7 | 8 |
| 67 | 2.559 | .570 | .581 | .591 | .602 | .613 | .624 | .635 | .647 | .658 | 2 | 4 | 6 | 7 | 9 |
| 68 | 2.669 | .681 | .693 | .705 | .716 | .729 | .741 | .753 | .765 | .778 | 2 | 4 | 6 | 8 | 10 |
| 69 | 2.790 | .803 | .816 | .829 | .842 | .855 | .869 | .882 | .896 | .910 | 2 | 4 | 7 | 9 | 11 |
| 70 | 2.924 | .938 | .952 | .967 | .981 | 2.996 | 3.011 | .026 | .041 | .056 | 2 | 5 | 7 | 10 | 12 |
| 71 | 3.072 | .087 | .103 | .119 | .135 | .152 | .168 | .185 | .202 | .219 | 3 | 5 | 8 | 11 | 14 |
| 72 | 3.236 | .254 | .271 | .289 | .307 | .326 | .344 | .363 | .382 | .401 | 3 | 6 | 9 | 12 | 15 |
| 73 | 3.420 | .440 | .460 | .480 | .500 | .521 | .542 | .563 | .584 | .606 | 3 | 7 | 10 | 14 | 17 |
| 74 | 3.628 | .650 | .673 | .695 | .719 | .742 | .766 | .790 | .814 | .839 | 4 | 8 | 12 | 16 | 20 |
| 75 | 3.864 | .889 | .915 | .941 | .967 | 3.994 | 4.021 | .049 | .077 | .105 | 4 | 9 | 14 | 18 | 22 |
| 76 | 4.134 | .163 | .192 | .222 | .253 | .284 | .315 | .347 | .379 | .412 | 5 | 10 | 16 | 21 | 26 |
| 77 | 4.445 | .479 | .514 | .549 | .584 | .620 | .657 | .694 | .732 | .771 | 6 | 12 | 18 | 24 | 30 |
| 78 | 4.810 | .850 | .890 | .931 | 4.973 | 5.016 | .059 | .103 | .148 | .194 | 7 | 14 | 22 | 29 | 36 |
| 79 | 5.241 | .288 | .337 | .386 | .436 | .487 | .540 | .593 | .647 | .702 | 9 | 17 | 26 | 35 | 43 |
| 80 | 5.759 | .816 | .875 | .935 | 5.996 | 6.059 | .123 | .188 | .255 | .323 | p.p. cease to be sufficiently accurate | | | | |
| 81 | 6.392 | .464 | .537 | .611 | .687 | .765 | .845 | 6.927 | 7.011 | .097 | | | | | |
| 82 | 7.185 | .276 | .368 | .463 | .561 | .661 | .764 | .870 | 7.979 | 8.091 | | | | | |
| 83 | 8.206 | .324 | .446 | .571 | .700 | .834 | 8.971 | 9.113 | .259 | .411 | | | | | |
| 84 | 9.567 | .728 | 9.895 | 10.068 | .248 | .433 | .626 | 10.826 | 11.034 | .249 | | | | | |
| 85 | 11.47 | 11.71 | 11.95 | 12.20 | 12.47 | 12.75 | 13.03 | 13.34 | 13.65 | 13.99 | | | | | |
| 86 | 14.34 | 14.70 | 15.09 | 15.50 | 15.93 | 16.38 | 16.86 | 17.37 | 17.91 | 18.49 | | | | | |
| 87 | 19.11 | 19.77 | 20.47 | 21.23 | 22.04 | 22.93 | 23.88 | 24.92 | 26.05 | 27.29 | | | | | |
| 88 | 28.65 | 30.16 | 31.84 | 33.71 | 35.81 | 38.20 | 40.93 | 44.08 | 47.75 | 52.09 | | | | | |
| 89 | 57.30 | 63.66 | 71.62 | 81.85 | 95.49 | 114.6 | 143.2 | 191.0 | 286.5 | 573.0 | | | | | |
| | 0′ | 6′ | 12′ | 18′ | 24′ | 30′ | 36′ | 42′ | 48′ | 54′ | 1′ | 2′ | 3′ | 4′ | 5′ |

# NATURAL TANGENTS

Proportional Parts

| | 0′ | 6′<br>0.1 | 12′<br>0.2 | 18′<br>0.3 | 24′<br>0.4 | 30′<br>0.5 | 36′<br>0.6 | 42′<br>0.7 | 48′<br>0.8 | 54′<br>0.9 | 1′ | 2′ | 3′ | 4′ | 5′ |
|---|---|---|---|---|---|---|---|---|---|---|---|---|---|---|---|
| 0° | 0.0000 | .0017 | .0035 | .0052 | .0070 | .0087 | .0105 | .0122 | .0140 | .0157 | 3 | 6 | 9 | 12 | 15 |
| 1 | 0.0175 | .0192 | .0209 | .0227 | .0244 | .0262 | .0279 | .0297 | .0314 | .0332 | 3 | 6 | 9 | 12 | 15 |
| 2 | 0.0349 | .0367 | .0384 | .0402 | .0419 | .0437 | .0454 | .0472 | .0489 | .0507 | 3 | 6 | 9 | 12 | 15 |
| 3 | 0.0524 | .0542 | .0559 | .0577 | .0594 | .0612 | .0629 | .0647 | .0664 | .0682 | 3 | 6 | 9 | 12 | 15 |
| 4 | 0.0699 | .0717 | .0734 | .0752 | .0769 | .0787 | .0805 | .0822 | .0840 | .0857 | 3 | 6 | 9 | 12 | 15 |
| 5 | 0.0875 | .0892 | .0910 | .0928 | .0945 | .0963 | .0981 | .0998 | .1016 | .1033 | 3 | 6 | 9 | 12 | 15 |
| 6 | 0.1051 | .1069 | .1086 | .1104 | .1122 | .1139 | .1157 | .1175 | .1192 | .1210 | 3 | 6 | 9 | 12 | 15 |
| 7 | 0.1228 | .1146 | .1263 | .1281 | .1299 | .1317 | .1334 | .1352 | .1370 | .1388 | 3 | 6 | 9 | 12 | 15 |
| 8 | 0.1405 | .1423 | .1441 | .1459 | .1477 | .1495 | .1512 | .1530 | .1548 | .1566 | 3 | 6 | 9 | 12 | 15 |
| 9 | 0.1584 | .1602 | .1620 | .1638 | .1655 | .1673 | .1691 | .1709 | .1727 | .1745 | 3 | 6 | 9 | 12 | 15 |
| 10 | 0.1763 | .1781 | .1799 | .1817 | .1835 | .1853 | .1871 | .1890 | .1908 | .1926 | 3 | 6 | 9 | 12 | 15 |
| 11 | 0.1944 | .1962 | .1980 | .1998 | .2016 | .2035 | .2053 | .2071 | .2089 | .2107 | 3 | 6 | 9 | 12 | 15 |
| 12 | 0.2126 | .2144 | .2162 | .2180 | .2199 | .2217 | .2235 | .2254 | .2272 | .2290 | 3 | 6 | 9 | 12 | 15 |
| 13 | 0.2309 | .2327 | .2345 | .2364 | .2382 | .2401 | .2419 | .2438 | .2456 | .2475 | 3 | 6 | 9 | 12 | 15 |
| 14 | 0.2493 | .2512 | .2530 | .2549 | .2568 | .2586 | .2605 | .2623 | .2642 | .2661 | 3 | 6 | 9 | 12 | 16 |
| 15 | 0.2679 | .2698 | .2717 | .2736 | .2754 | .2773 | .2792 | .2811 | .2830 | .2849 | 3 | 6 | 9 | 13 | 16 |
| 16 | 0.2867 | .2886 | .2905 | .2924 | .2943 | .2962 | .2981 | .3000 | .3019 | .3038 | 3 | 6 | 9 | 13 | 16 |
| 17 | 0.3057 | .3076 | .3096 | .3115 | .3134 | .3153 | .3172 | .3191 | .3211 | .3230 | 3 | 6 | 9 | 13 | 16 |
| 18 | 0.3249 | .3269 | .3288 | .3307 | .3327 | .3346 | .3365 | .3385 | .3404 | .3424 | 3 | 6 | 10 | 13 | 16 |
| 19 | 0.3443 | .3463 | .3482 | .3502 | .3522 | .3541 | .3561 | .3581 | .3600 | .3620 | 3 | 6 | 10 | 13 | 16 |
| 20 | 0.3640 | .3659 | .3679 | .3699 | .3719 | .3739 | .3759 | .3779 | .3799 | .3819 | 3 | 6 | 10 | 13 | 17 |
| 21 | 0.3839 | .3859 | .3879 | .3899 | .3919 | .3939 | .3959 | .3979 | .4000 | .4020 | 3 | 7 | 10 | 13 | 17 |
| 22 | 0.4040 | .4061 | .4081 | .4101 | .4122 | .4142 | .4163 | .4183 | .4204 | .4224 | 3 | 7 | 10 | 14 | 17 |
| 23 | 0.4245 | .4265 | .4286 | .4307 | .4327 | .4348 | .4369 | .4390 | .4411 | .4431 | 3 | 7 | 10 | 14 | 17 |
| 24 | 0.4452 | .4473 | .4494 | .4515 | .4536 | .4557 | .4578 | .4599 | .4621 | .4642 | 4 | 7 | 11 | 14 | 18 |
| 25 | 0.4663 | .4684 | .4706 | .4727 | .4748 | .4770 | .4791 | .4813 | .4834 | .4856 | 4 | 7 | 11 | 14 | 18 |
| 26 | 0.4877 | .4899 | .4921 | .4942 | .4964 | .4986 | .5008 | .5029 | .5051 | .5073 | 4 | 7 | 11 | 15 | 18 |
| 27 | 0.5095 | .5117 | .5139 | .5161 | .5184 | .5206 | .5228 | .5250 | .5272 | .5295 | 4 | 7 | 11 | 15 | 18 |
| 28 | 0.5317 | .5339 | .5362 | .5384 | .5407 | .5430 | .5452 | .5475 | .5498 | .5520 | 4 | 8 | 11 | 15 | 19 |
| 29 | 0.5543 | .5566 | .5589 | .5612 | .5635 | .5658 | .5681 | .5704 | .5727 | .5750 | 4 | 8 | 12 | 15 | 19 |
| 30 | 0.5774 | .5797 | .5820 | .5844 | .5867 | .5891 | .5914 | .5938 | .5961 | .5985 | 4 | 8 | 12 | 16 | 20 |
| 31 | 0.6009 | .6032 | .6056 | .6080 | .6104 | .6128 | .6152 | .6176 | .6200 | .6224 | 4 | 8 | 12 | 16 | 20 |
| 32 | 0.6249 | .6273 | .6297 | .6322 | .6346 | .6371 | .6395 | .6420 | .6445 | .6469 | 4 | 8 | 12 | 16 | 20 |
| 33 | 0.6494 | .6519 | .6544 | .6569 | .6594 | .6619 | .6644 | .6669 | .6694 | .6720 | 4 | 8 | 13 | 17 | 21 |
| 34 | 0.6745 | .6771 | .6796 | .6822 | .6847 | .6873 | .6899 | .6924 | .6950 | .6976 | 4 | 9 | 13 | 17 | 21 |
| 35 | 0.7002 | .7028 | .7054 | .7080 | .7107 | .7133 | .7159 | .7186 | .7212 | .7239 | 4 | 9 | 13 | 18 | 22 |
| 36 | 0.7265 | .7292 | .7319 | .7346 | .7373 | .7400 | .7427 | .7454 | .7481 | .7508 | 5 | 9 | 14 | 18 | 23 |
| 37 | 0.7536 | .7563 | .7590 | .7618 | .7646 | .7673 | .7701 | .7729 | .7757 | .7785 | 5 | 9 | 14 | 18 | 23 |
| 38 | 0.7813 | .7841 | .7869 | .7898 | .7926 | .7954 | .7983 | .8012 | .8040 | .8069 | 5 | 10 | 14 | 19 | 24 |
| 39 | 0.8098 | .8127 | .8156 | .8185 | .8214 | .8243 | .8273 | .8302 | .8332 | .8361 | 5 | 10 | 15 | 20 | 24 |
| 40 | 0.8391 | .8421 | .8451 | .8481 | .8511 | .8541 | .8571 | .8601 | .8632 | .8662 | 5 | 10 | 15 | 20 | 25 |
| 41 | 0.8693 | .8724 | .8754 | .8785 | .8816 | .8847 | .8878 | .8910 | .8941 | .8972 | 5 | 10 | 16 | 21 | 26 |
| 42 | 0.9004 | .9036 | .9067 | .9099 | .9131 | .9163 | .9195 | .9228 | .9260 | .9293 | 5 | 11 | 16 | 21 | 26 |
| 43 | 0.9325 | .9358 | .9391 | .9424 | .9457 | .9490 | .9523 | .9556 | .9590 | .9623 | 6 | 11 | 17 | 22 | 28 |
| 44 | 0.9657 | .9691 | .9725 | .9759 | .9793 | .9827 | .9861 | .9896 | .9930 | .9965 | 6 | 11 | 17 | 23 | 29 |
| | 0′ | 6′ | 12′ | 18′ | 24′ | 30′ | 36′ | 42′ | 48′ | 54′ | 1′ | 2′ | 3′ | 4′ | 5′ |

# NATURAL TANGENTS

Proportional Parts

| | 0′ | 6′ 0.1 | 12′ 0.2 | 18′ 0.3 | 24′ 0.4 | 30′ 0.5 | 36′ 0.6 | 42′ 0.7 | 48′ 0.8 | 54′ 0.9 | 1′ | 2′ | 3′ | 4′ | 5′ |
|---|---|---|---|---|---|---|---|---|---|---|---|---|---|---|---|
| 45° | 1.0000 | .0035 | .0070 | .0105 | .0141 | .0176 | .0212 | .0247 | .0283 | .0319 | 6 | 12 | 18 | 24 | 30 |
| 46 | 1.0355 | .0392 | .0428 | .0464 | .0501 | .0538 | .0575 | .0612 | .0649 | .0686 | 6 | 12 | 18 | 25 | 31 |
| 47 | 1.0724 | .0761 | .0799 | .0837 | .0875 | .0913 | .0951 | .0990 | .1028 | .1067 | 6 | 13 | 19 | 25 | 32 |
| 48 | 1.1106 | .1145 | .1184 | .1224 | .1263 | .1303 | .1343 | .1383 | .1423 | .1463 | 7 | 13 | 20 | 27 | 33 |
| 49 | 1.1504 | .1544 | .1585 | .1626 | .1667 | .1708 | .1750 | .1792 | .1833 | .1875 | 7 | 14 | 21 | 28 | 34 |
| 50 | 1.1918 | .1960 | .2002 | .2045 | .2088 | .2131 | .2174 | .2218 | .2261 | .2305 | 7 | 14 | 22 | 29 | 36 |
| 51 | 1.2349 | .2393 | .2437 | .2482 | .2527 | .2572 | .2617 | .2662 | .2708 | .2753 | 8 | 15 | 23 | 30 | 38 |
| 52 | 1.2799 | .2846 | .2892 | .2938 | .2985 | .3032 | .3079 | .3127 | .3175 | .3222 | 8 | 16 | 24 | 31 | 39 |
| 53 | 1.3270 | .3319 | .3367 | .3416 | .3465 | .3514 | .3564 | .3613 | .3663 | .3713 | 8 | 16 | 25 | 33 | 41 |
| 54 | 1.3764 | .3814 | .3865 | .3916 | .3968 | .4019 | .4071 | .4124 | .4176 | .3229 | 9 | 17 | 26 | 34 | 43 |
| 55 | 1.4281 | .4335 | .4383 | .4442 | .4496 | .4550 | .4605 | .4659 | .4715 | .4770 | 9 | 18 | 27 | 36 | 45 |
| 56 | 1.4826 | .4882 | .4938 | .4994 | .5051 | .5108 | .5166 | .5224 | .5282 | .5340 | 10 | 19 | 29 | 38 | 48 |
| 57 | 1.5399 | .5458 | .5517 | .5577 | .5637 | .5697 | .5757 | .5818 | .5880 | .5941 | 10 | 20 | 30 | 40 | 50 |
| 58 | 1.6003 | .6066 | .6128 | .6191 | .6255 | .6319 | .6383 | .6447 | .6512 | .6577 | 11 | 21 | 32 | 43 | 53 |
| 59 | 1.6643 | .6709 | .6775 | .6842 | .6909 | .6977 | .7045 | .7113 | .7182 | .7251 | 11 | 23 | 34 | 45 | 57 |
| 60 | 1.7321 | .7391 | .7461 | .7532 | .7603 | .7675 | .7747 | .7820 | .7893 | .7966 | 12 | 24 | 36 | 48 | 60 |
| 61 | 1.8040 | .8115 | .8190 | .8265 | .8341 | .8418 | .8495 | .8572 | .8650 | .8728 | 13 | 26 | 38 | 51 | 64 |
| 62 | 1.8807 | .8887 | .8967 | .9047 | .9128 | .9210 | .9292 | .9375 | .9458 | .9542 | 14 | 27 | 41 | 55 | 68 |
| 63 | 1.9626 | .9711 | .9797 | .9883 | 1.9970 | 2.0057 | .0145 | .0233 | .0323 | .0413 | 15 | 29 | 44 | 58 | 73 |
| 64 | 2.0503 | .0594 | .0686 | .0778 | .0872 | .0965 | .1060 | .1155 | .1251 | .1348 | 16 | 31 | 47 | 63 | 78 |
| 65 | 2.145 | .154 | .164 | .174 | .184 | .194 | .204 | .215 | .225 | .236 | 2 | 3 | 5 | 7 | 8 |
| 66 | 2.246 | .257 | .267 | .278 | .289 | .300 | .311 | .322 | .333 | .344 | 2 | 4 | 5 | 7 | 9 |
| 67 | 2.356 | .367 | .379 | .391 | .402 | .414 | .426 | .438 | .450 | .463 | 2 | 4 | 6 | 8 | 10 |
| 68 | 2.475 | .488 | .500 | .513 | .526 | .539 | .552 | .565 | .578 | .592 | 2 | 4 | 6 | 9 | 11 |
| 69 | 2.605 | .619 | .633 | .646 | .660 | .675 | .689 | .703 | .718 | .733 | 2 | 5 | 7 | 9 | 12 |
| 70 | 2.747 | .762 | .778 | .793 | .808 | .824 | .840 | .856 | .872 | .888 | 3 | 5 | 8 | 10 | 13 |
| 71 | 2.904 | .921 | .937 | .954 | .971 | 2.989 | 3.006 | .024 | .042 | .060 | 3 | 6 | 9 | 12 | 14 |
| 72 | 3.078 | .096 | .115 | .133 | .152 | .172 | .191 | .211 | .230 | .251 | 3 | 6 | 10 | 13 | 16 |
| 73 | 3.271 | .291 | .312 | .333 | .354 | .376 | .398 | .420 | .442 | .465 | 4 | 7 | 11 | 14 | 18 |
| 74 | 3.487 | .511 | .534 | .558 | .582 | .606 | .630 | .655 | .681 | .706 | 4 | 8 | 12 | 16 | 20 |
| 75 | 3.732 | .758 | .785 | .812 | .839 | .867 | .895 | .923 | .952 | .981 | 5 | 9 | 14 | 19 | 23 |
| 76 | 4.011 | .041 | .071 | .102 | .134 | .165 | .198 | .230 | .264 | .297 | 5 | 11 | 16 | 21 | 27 |
| 77 | 4.331 | .366 | .402 | .437 | .474 | .511 | .548 | .586 | .625 | .665 | 6 | 12 | 19 | 25 | 31 |
| 78 | 4.705 | .745 | .787 | .829 | .872 | .915 | 4.959 | 5.005 | .050 | .097 | 7 | 15 | 22 | 29 | 37 |
| 79 | 5.145 | .193 | .242 | .292 | .343 | .396 | .449 | .503 | .558 | .614 | 9 | 18 | 26 | 35 | 44 |
| 80 | 5.671 | .730 | .789 | .850 | .912 | 5.976 | 6.041 | .107 | .174 | .243 | 11 | 21 | 32 | 43 | 54 |
| 81 | 6.314 | .386 | .460 | .535 | .612 | .691 | .772 | .855 | 6.940 | 7.026 | 13 | 27 | 40 | 54 | 67 |
| 82 | 7.115 | .207 | .300 | .396 | .495 | .596 | .700 | .806 | 7.916 | 8.028 | 17 | 34 | 51 | 69 | 86 |
| 83 | 8.144 | .264 | .386 | .513 | .643 | .777 | 8.915 | 9.058 | .205 | .357 | 23 | 46 | 68 | 91 | 114 |
| 84 | 9.514 | 9.677 | 9.845 | 10.019 | 10.199 | 10.385 | 10.579 | 10.780 | 10.988 | 11.205 | | | | | |
| 85 | 11.43 | 11.66 | 11.91 | 12.16 | 12.43 | 12.71 | 13.00 | 13.30 | 13.62 | 13.95 | p.p. cease to be sufficiently accurate | | | | |
| 86 | 14.30 | 14.67 | 15.06 | 15.46 | 15.89 | 16.35 | 16.83 | 17.34 | 17.89 | 18.46 | | | | | |
| 87 | 19.08 | 19.74 | 20.45 | 21.20 | 22.02 | 22.90 | 23.86 | 24.90 | 26.03 | 27.27 | | | | | |
| 88 | 28.64 | 30.14 | 31.82 | 33.69 | 35.80 | 38.19 | 40.92 | 44.07 | 47.74 | 52.08 | | | | | |
| 89 | 57.29 | 63.66 | 71.62 | 81.85 | 95.49 | 114.6 | 143.2 | 191.0 | 286.5 | 573.0 | | | | | |
| | 0′ | 6′ | 12′ | 18′ | 24′ | 30′ | 36′ | 42′ | 48′ | 54′ | 1′ | 2′ | 3′ | 4′ | 5′ |

# NATURAL COTANGENTS

Proportional Parts Subtract

|  | 0′ | 6′ 0.1 | 12′ 0.2 | 18′ 0.3 | 24′ 0.4 | 30′ 0.5 | 36′ 0.6 | 42′ 0.7 | 48′ 0.8 | 54′ 0.9 | 1′ | 2′ | 3′ | 4′ | 5′ |
|---|---|---|---|---|---|---|---|---|---|---|---|---|---|---|---|
| 0° | ∞ | 573.0 | 286.5 | 191.0 | 143.2 | 114.6 | 95.49 | 81.85 | 71.62 | 63.66 | p.p. cease to be sufficiently accurate | | | | |
| 1 | 57.29 | 52.08 | 47.74 | 44.07 | 40.92 | 38.19 | 35.80 | 33.69 | 31.82 | 30.14 | | | | | |
| 2 | 28.64 | 27.27 | 26.03 | 24.90 | 23.86 | 22.90 | 22.02 | 21.20 | 20.45 | 19.74 | | | | | |
| 3 | 19.08 | 18.46 | 17.89 | 17.34 | 16.83 | 16.35 | 15.89 | 15.46 | 15.06 | 14.67 | | | | | |
| 4 | 14.30 | 13.95 | 13.62 | 13.30 | 13.00 | 12.71 | 12.43 | 12.16 | 11.91 | 11.66 | | | | | |
| 5 | 11.30 | 11.205 | 10.988 | 10.780 | 10.579 | 10.385 | 10.199 | 10.019 | 9.845 | 9.677 | | | | | |
| 6 | 9.514 | .357 | .205 | 9.058 | 8.915 | .777 | .643 | .513 | .386 | .264 | 23 | 46 | 68 | 91 | 114 |
| 7 | 8.144 | 8.028 | 7.916 | .806 | .700 | .596 | .495 | .396 | .300 | .207 | 17 | 34 | 51 | 69 | 86 |
| 8 | 7.115 | 7.026 | 6.940 | .855 | .772 | .691 | .612 | .535 | .460 | .386 | 13 | 27 | 40 | 54 | 67 |
| 9 | 6.314 | .243 | .174 | .107 | 6.041 | 5.976 | .912 | .850 | .789 | .730 | 11 | 21 | 32 | 43 | 54 |
| 10 | 5.671 | .614 | .588 | .503 | .449 | .396 | .343 | .292 | .242 | .193 | 9 | 18 | 26 | 35 | 44 |
| 11 | 5.145 | .097 | .050 | 5.005 | 4.959 | .915 | .872 | .829 | .787 | .745 | 7 | 15 | 22 | 29 | 37 |
| 12 | 4.705 | .665 | .625 | .586 | .548 | .511 | .474 | .427 | .402 | .366 | 6 | 12 | 19 | 25 | 31 |
| 13 | 4.331 | .297 | .264 | .230 | .198 | .165 | .134 | .102 | .071 | .041 | 5 | 11 | 16 | 21 | 27 |
| 14 | 4.011 | 3.981 | .952 | .923 | .895 | .867 | .839 | .812 | .785 | .758 | 5 | 9 | 14 | 19 | 23 |
| 15 | 3.732 | .706 | .681 | .655 | .630 | .606 | .582 | .558 | .534 | .511 | 4 | 8 | 12 | 16 | 20 |
| 16 | 3.487 | .465 | .442 | .420 | .398 | .376 | .354 | .333 | .312 | .291 | 4 | 7 | 11 | 14 | 18 |
| 17 | 3.271 | .251 | .230 | .211 | .191 | .172 | .152 | .133 | .115 | .096 | 3 | 6 | 10 | 13 | 16 |
| 18 | 3.078 | .060 | .042 | .024 | 3.006 | 2.989 | .071 | .954 | .937 | .921 | 3 | 6 | 9 | 12 | 14 |
| 19 | 2.904 | .888 | .872 | .856 | .840 | .824 | .808 | .793 | .778 | .762 | 3 | 5 | 8 | 10 | 13 |
| 20 | 2.747 | .733 | .718 | .703 | .689 | .675 | .660 | .646 | .633 | .619 | 2 | 5 | 7 | 9 | 12 |
| 21 | 2.605 | .592 | .578 | .565 | .552 | .539 | .526 | .513 | .500 | .488 | 2 | 4 | 6 | 9 | 11 |
| 22 | 2.475 | .463 | .450 | .438 | .426 | .414 | .402 | .391 | .379 | .367 | 2 | 4 | 6 | 8 | 10 |
| 23 | 2.356 | .344 | .333 | .322 | .311 | .300 | .289 | .278 | .267 | .257 | 2 | 4 | 5 | 7 | 9 |
| 24 | 2.246 | .236 | .225 | .215 | .204 | .194 | .184 | .174 | .164 | .154 | 2 | 3 | 5 | 7 | 8 |
| 25 | 2.1445 | .1348 | .1251 | .1155 | .1060 | .0965 | .0872 | .0778 | .0686 | .0594 | 16 | 31 | 47 | 63 | 78 |
| 26 | 2.0503 | .0413 | .0323 | .0233 | .0145 | 2.0057 | 1.9970 | .09883 | .9797 | .9711 | 15 | 29 | 44 | 58 | 73 |
| 27 | 1.9626 | .9542 | .9548 | .9375 | .9292 | .9210 | .9128 | .9047 | .8967 | .8887 | 14 | 27 | 41 | 55 | 68 |
| 28 | 1.8807 | .8728 | .8650 | .8572 | .8495 | .8418 | .8341 | .8265 | .8190 | .8115 | 13 | 26 | 38 | 51 | 64 |
| 29 | 1.8040 | .7966 | .7893 | .7820 | .7747 | .7675 | .7603 | .7532 | .7461 | .7391 | 12 | 24 | 36 | 48 | 60 |
| 30 | 1.7321 | .7251 | .7182 | .7113 | .7045 | .6977 | .6909 | .6842 | .6775 | .6709 | 11 | 23 | 34 | 45 | 57 |
| 31 | 1.6643 | .6577 | .6512 | .6447 | .6383 | .6319 | .6255 | .6191 | .6128 | .6066 | 11 | 21 | 32 | 43 | 53 |
| 32 | 1.6003 | .5941 | .5880 | .5818 | .5757 | .5697 | .5637 | .5577 | .5517 | .5458 | 10 | 20 | 30 | 40 | 50 |
| 33 | 1.5399 | .5340 | .5282 | .5224 | .5166 | .5108 | .5051 | .4994 | .4938 | .4882 | 10 | 19 | 29 | 38 | 48 |
| 34 | 1.4826 | .4770 | .4715 | .4659 | .4605 | .4550 | .4496 | .4442 | .4388 | .4335 | 9 | 18 | 27 | 36 | 45 |
| 35 | 1.4281 | .4229 | .4176 | .4124 | .4071 | .4019 | .3968 | .3916 | .3865 | .3814 | 9 | 17 | 26 | 34 | 43 |
| 36 | 1.3764 | .3713 | .3663 | .3613 | .3564 | .3514 | .3465 | .3416 | .3367 | .3319 | 8 | 16 | 25 | 33 | 41 |
| 37 | 1.3270 | .3222 | .3175 | .3127 | .3079 | .3032 | .2985 | .2938 | .2892 | .2846 | 8 | 16 | 24 | 31 | 39 |
| 38 | 1.2799 | .2753 | .2708 | .2662 | .2617 | .2572 | .2527 | .2482 | .2437 | .2393 | 8 | 15 | 23 | 30 | 38 |
| 39 | 1.2349 | .2305 | .2261 | .2218 | .2174 | .2131 | .2088 | .2045 | .2002 | .1960 | 7 | 14 | 22 | 29 | 36 |
| 40 | 1.1918 | .1875 | .1833 | .1792 | .1750 | .1708 | .1667 | .1626 | .1585 | .1544 | 7 | 14 | 21 | 28 | 34 |
| 41 | 1.1504 | .1463 | .1423 | .1383 | .1343 | .1303 | .1263 | .1224 | .1184 | .1145 | 7 | 13 | 20 | 27 | 33 |
| 42 | 1.1106 | .1067 | .1028 | .0990 | .0951 | .0913 | .0875 | .0837 | .0799 | .0761 | 6 | 13 | 19 | 25 | 32 |
| 43 | 1.0724 | .0686 | .0649 | .0612 | .0575 | .0538 | .0501 | .0464 | .0428 | .0392 | 6 | 12 | 18 | 25 | 31 |
| 44 | 1.0355 | .0319 | .0283 | .0247 | .0212 | .0176 | .0141 | .0105 | .0070 | .0035 | 6 | 12 | 18 | 24 | 30 |
|  | 0′ | 6′ | 12′ | 18′ | 24′ | 30′ | 36′ | 42′ | 48′ | 54′ | 1′ | 2′ | 3′ | 4′ | 5′ |

# NATURAL COTANGENTS

Proportional Parts Subtract

| | 0′ | 6′ 0.1 | 12′ 0.2 | 18′ 0.3 | 24′ 0.4 | 30′ 0.5 | 36′ 0.6 | 42′ 0.7 | 48′ 0.8 | 54′ 0.9 | 1′ | 2′ | 3′ | 4′ | 5′ |
|---|---|---|---|---|---|---|---|---|---|---|---|---|---|---|---|
| 45° | 1.0000 | 0.9965 | .9930 | .9896 | .9861 | .9827 | .9793 | .9759 | .9725 | .9691 | 6 | 11 | 17 | 23 | 29 |
| 46 | 0.9657 | .9623 | .9590 | .9556 | .9523 | .9490 | .9457 | .9424 | .9391 | .9358 | 6 | 11 | 17 | 22 | 28 |
| 47 | 0.9325 | .9293 | .9260 | .9228 | .9195 | .9163 | .9131 | .9099 | .9067 | .9036 | 5 | 11 | 16 | 21 | 27 |
| 48 | 0.9004 | .8972 | .8941 | .8910 | .8878 | .8847 | .8816 | .8785 | .8754 | .8724 | 5 | 10 | 16 | 21 | 26 |
| 49 | 0.8693 | .8662 | .8632 | .8601 | .8571 | .8541 | .8511 | .8481 | .8451 | .8421 | 5 | 10 | 15 | 20 | 25 |
| 50 | 0.8391 | .8361 | .8332 | .8302 | .8273 | .8243 | .8214 | .8185 | .8156 | .8127 | 5 | 10 | 15 | 20 | 24 |
| 51 | 0.8098 | .8069 | .8040 | .8012 | .7983 | .7954 | .7926 | .7898 | .7869 | .7841 | 5 | 10 | 14 | 19 | 24 |
| 52 | 0.7813 | .7785 | .7757 | .7729 | .7701 | .7673 | .7646 | .7618 | .7590 | .7563 | 5 | 9 | 14 | 18 | 23 |
| 53 | 0.7536 | .7508 | .7481 | .7454 | .7427 | .7400 | .7373 | .7346 | .7319 | .7292 | 5 | 9 | 14 | 18 | 23 |
| 54 | 0.7265 | .7239 | .7212 | .7186 | .7159 | .7133 | .7107 | .7080 | .7054 | .7028 | 4 | 9 | 13 | 18 | 22 |
| 55 | 0.7002 | .6976 | .6950 | .6924 | .6899 | .6873 | .6847 | .6822 | .6796 | .6771 | 4 | 9 | 13 | 17 | 21 |
| 56 | 0.6745 | .6720 | .6694 | .6669 | .6644 | .6519 | .6594 | .6569 | .6544 | .6519 | 4 | 8 | 13 | 17 | 21 |
| 57 | 0.6494 | .6469 | .6445 | .6420 | .6395 | .6371 | .6346 | .6322 | .6297 | .6273 | 4 | 8 | 12 | 16 | 20 |
| 58 | 0.6249 | .6224 | .6200 | .6176 | .6152 | .6128 | .6104 | .6080 | .6056 | .6032 | 4 | 8 | 12 | 16 | 20 |
| 59 | 0.6009 | .5985 | .5961 | .5938 | .5914 | .5891 | .5867 | .5844 | .5820 | .5797 | 4 | 8 | 12 | 16 | 20 |
| 60 | 0.5774 | .5750 | .5727 | .5704 | .5681 | .5658 | .5635 | .5612 | .5589 | .5566 | 4 | 8 | 12 | 15 | 19 |
| 61 | 0.5543 | .5520 | .5498 | .5475 | .5452 | .5430 | .5407 | .5384 | .5362 | .5339 | 4 | 8 | 11 | 15 | 19 |
| 62 | 0.5317 | .5295 | .5272 | .5250 | .5228 | .5206 | .5184 | .5161 | .5139 | .5117 | 4 | 7 | 11 | 15 | 18 |
| 63 | 0.5095 | .5073 | .5051 | .5029 | .5008 | .4986 | .4964 | .4942 | .4921 | .4899 | 4 | 7 | 11 | 15 | 18 |
| 64 | 0.4877 | .4856 | .4834 | .4813 | .4791 | .4770 | .4748 | .4727 | .4706 | .4684 | 4 | 7 | 11 | 14 | 18 |
| 65 | 0.4663 | .4642 | .4621 | .4599 | .4578 | .4557 | .4536 | .4515 | .4494 | .4473 | 4 | 7 | 11 | 14 | 18 |
| 66 | 0.4452 | .4431 | .4411 | .4390 | .4369 | .4348 | .4327 | .4307 | .4286 | .4265 | 3 | 7 | 10 | 14 | 17 |
| 67 | 0.4245 | .4224 | .4204 | .4183 | .4163 | .4142 | .4122 | .4101 | .4081 | .4061 | 3 | 7 | 10 | 14 | 17 |
| 68 | 0.4040 | .4020 | .4000 | .3979 | .3959 | .3939 | .3919 | .3899 | .3879 | .3859 | 3 | 7 | 10 | 13 | 17 |
| 69 | 0.3839 | .3819 | .3799 | .3779 | .3759 | .3739 | .3719 | .3699 | .3679 | .3659 | 3 | 6 | 10 | 13 | 17 |
| 70 | 0.3640 | .3620 | .3600 | .3581 | .3561 | .3541 | .3522 | .3502 | .3482 | .3463 | 3 | 6 | 10 | 13 | 16 |
| 71 | 0.3443 | .3424 | .3404 | .3385 | .3365 | .3346 | .3327 | .3307 | .3288 | .3269 | 3 | 6 | 10 | 13 | 16 |
| 72 | 0.3249 | .3230 | .3211 | .3191 | .3172 | .3153 | .3134 | .3115 | .3096 | .3076 | 3 | 6 | 9 | 13 | 16 |
| 73 | 0.3057 | .3038 | .3019 | .3000 | .2981 | .2962 | .2943 | .2924 | .2905 | .2886 | 3 | 6 | 9 | 13 | 16 |
| 74 | 0.2867 | .2849 | .2830 | .2811 | .2792 | .2773 | .2754 | .2736 | .2717 | .2698 | 3 | 6 | 9 | 13 | 16 |
| 75 | 0.2679 | .2661 | .2642 | .2623 | .2605 | .2586 | .2568 | .2549 | .2530 | .2512 | 3 | 6 | 9 | 12 | 16 |
| 76 | 0.2493 | .2475 | .2456 | .2438 | .2419 | .2401 | .2382 | .2364 | .2345 | .2327 | 3 | 6 | 9 | 12 | 15 |
| 77 | 0.2309 | .2290 | .2272 | .2254 | .2235 | .2217 | .2199 | .2180 | .2162 | .2144 | 3 | 6 | 9 | 12 | 15 |
| 78 | 0.2126 | .2107 | .2089 | .2071 | .2053 | .2035 | .2016 | .1998 | .1980 | .1962 | 3 | 6 | 9 | 12 | 15 |
| 79 | 0.1944 | .1926 | .1908 | .1890 | .1871 | .1853 | .1835 | .1817 | .1799 | .1781 | 3 | 6 | 9 | 12 | 15 |
| 80 | 0.1763 | .1745 | .1727 | .1709 | .1691 | .1673 | .1655 | .1638 | .1620 | .1602 | 3 | 6 | 9 | 12 | 15 |
| 81 | 0.1584 | .1566 | .1548 | .1530 | .1512 | .1495 | .1477 | .1459 | .1441 | .1423 | 3 | 6 | 9 | 12 | 15 |
| 82 | 0.1405 | .1388 | .1370 | .1352 | .1334 | .1317 | .1299 | .1281 | .1263 | .1246 | 3 | 6 | 9 | 12 | 15 |
| 83 | 0.1228 | .1210 | .1192 | .1175 | .1257 | .1139 | .1122 | .1104 | .1086 | .1069 | 3 | 6 | 9 | 12 | 15 |
| 84 | 0.1051 | .1033 | .1016 | .0998 | .0981 | .0963 | .0945 | .0928 | .0910 | .0892 | 3 | 6 | 9 | 12 | 15 |
| 85 | 0.0875 | .0857 | .0840 | .0822 | .0805 | .0787 | .0769 | .0752 | .0734 | .0717 | 3 | 6 | 9 | 12 | 15 |
| 86 | 0.0699 | .0682 | .0664 | .0647 | .0629 | .0612 | .0594 | .0577 | .0559 | .0542 | 3 | 6 | 9 | 12 | 15 |
| 87 | 0.0524 | .0507 | .0489 | .0472 | .0454 | .0437 | .0419 | .0402 | .0384 | .0367 | 3 | 6 | 9 | 12 | 15 |
| 88 | 0.0349 | .0332 | .0314 | .0297 | .0279 | .0262 | .0244 | .0227 | .0209 | .0192 | 3 | 6 | 9 | 12 | 15 |
| 89 | 0.0175 | .0157 | .0140 | .0122 | .0105 | .0087 | .0070 | .0052 | .0035 | .0017 | 3 | 6 | 9 | 12 | 15 |
| | 0′ | 6′ | 12′ | 18′ | 24′ | 30′ | 36′ | 42′ | 48′ | 54′ | 1′ | 2′ | 3′ | 4′ | 5′ |

# Answers

## Exercise 1 (p. 48)

**1** $\tan ABC = \frac{AC}{CB} = \frac{CD}{DB} = \frac{CQ}{QD} = \frac{DQ}{QB} = \frac{AD}{CD}$

$\tan CAB = \frac{CB}{AC} = \frac{DB}{CD} = \frac{QD}{CQ} = \frac{QB}{DQ} = \frac{CD}{AD}$

**2** $\tan ABC = \frac{4}{5}$, $\tan CAB = \frac{3}{4}$

**3** (1) 0.3249 (2) 0.9325 (3) 1.4826 (4) 3.2709 (5) 0.2549 (6) 0.6950

**4** (1) 0.1635 (2) 0.6188 (3) 0.8122 (4) 1.3009 (5) 2.1123

**5** (1) 28.6° (2) 61.3° (3) 70.5° (4) 52.43° (5) 33.85° (6) 14.27°

**6** 8.36 m **7** 67.38°, 67.38°, 45.24° **8** 19.54 m

**9** 1.41 km **10** 21.3 m approx. **11** 37°; 53° approx.

**12** 144 m

## Exercise 2 (p. 57)

**1** $\sin ABC = \frac{AC}{AB} = \frac{DQ}{DB} = \frac{CD}{CB} = \frac{CQ}{CD} = \frac{AD}{AC}$

$\sin CAB = \frac{CB}{AB} = \frac{QB}{DB} = \frac{DB}{CB} = \frac{DQ}{CD} = \frac{CD}{AC}$

$\cos ABC = \frac{CB}{AB} = \frac{QB}{DB} = \frac{DB}{CB} = \frac{DQ}{CD} = \frac{CD}{AC}$

$\cos CAB = \frac{AC}{AB} = \frac{DQ}{DB} = \frac{CD}{CB} = \frac{CQ}{CD} = \frac{AD}{AC}$

**2** Cosine is 0.1109, sine is 0.9939
**3** Length is 5.14 cm approx., distance from centre 3.06 cm approx.
**4** Sines 0.6 and 0.8, cosines 0.8 and 0.6
**5** (1) 0.2521 (2) 0.7400 (3) 0.9353
**6** (1) 29.8° (2) 30.77° (3) 52.23°
**7** (1) 0.9350 (2) 0.7149 (3) 0.4594 (4) 0.7789 (5) 0.1863 (6) 0.5390
**8** (1) 57.78° (2) 20.65° (3) 69.23° (4) 77.45° (5) 37.72° (6) 59.07°
**9** 10.08°
**10** 7.34 m; 37.8°; 52.2°
**11** 13.93°
**12** 47.6°; 43.8 m approx.

## Exercise 3 (p. 64)

**1** (1) 1.7263 (2) 1.1576 (3) 1.3589 (4) 1.6649 (5) 1.2045 (6) 0.3528
**2** (1) 60.62° (2) 64.75° (3) 69.3°
**3** 48.2 mm
**4** 22.62°, 67.38°
**5** 2.87 m
**6** 7.19 m
**7** (a) 0.3465 (b) 0.4394
**8** (a) 0.2204 (b) 2.988
**9** (a) 0.7357 (b) 1.691
**10** (a) 1.869 (b) 1.56 approx.
**11** 0.5602
**12** (1) 0.2616 (2) −0.4695
**13** 37.13°
**14** 1.2234
**15** 0.09661
**16** 553.5

## Exercise 4 (p. 71)

**1** 35.02°, 54.98°, 2.86 m
**2** 44.2°
**3** a = 55.5, b = 72.6
**4** A = 30.5°, B = 59.5°
**5** AD = 2.66 cm, BD = 1.87 cm, DC = 2.81 cm, AC = 3.87 cm
**6** A = 44.13°, b = 390 mm (approx.)
**7** 69.52°, 60°
**8** 10.3 km N., 14.7 km E.
**9** 0.68 cm
**10** $\frac{x\sqrt{3}}{2}$; $\frac{\sqrt{3}}{2}$, $\frac{1}{2}$

**11** 2.60 cm; 2.34 cm (both approx.)
**12** 3.6° **13** 10.2 km W., 11.7 km N
**14** 31.83° W. of N; 17.1 km

## Exercise 5 (p. 74)

**1** 0.7002 **2** $\frac{4}{5}$ ; $\frac{3}{4}$
**3** 0.8827 **4** 1.6243
**5** 0.6745, 0.8290, 0.5592 **6** 1.1547
**7** 1.9121; 0.5230; 0.8523

**8** $\sec\theta = \sqrt{1 + t^2}$; $\cos\theta = \dfrac{1}{\sqrt{1 + t^2}}$; $\sin\theta = \dfrac{t}{\sqrt{1 + t^2}}$

**9** $\sin\alpha = 0.8829$, $\tan\alpha = 1.8807$

## Exercise 6 (p. 85)

**1** sines are (a) 0.9781 (c) 0.9428 (e) 0.4289
(b) 0.5068 (d) 0.5698
cosines are (a) −0.2079 (c) −0.3333 (e) −0.9033
(b) −0.8621 (d) −0.8218
tangents are (a) −4.7046 (c) −2.8291 (e) −0.4748
(b) −0.5879 (d) −0.6933

**2** (a) 40.60° or 139.4° (c) 20.3° or 159.7°
(b) 65.87° or 114.13° (d) 45.42° or 134.58°

**3** (a) 117° (c) 100.3° (e) 142.35°
(b) 144.4° (d) 159.3° (f) 156.25°

**4** (a) 151° (c) 112.3° (e) 144.47°
(b) 123.8° (d) 119.6° (f) 130.38°

**5** (a) 2.2812 (b) −1.0485 c3) −3.3122

**6** (a) 127.27° (d) 24° or 156°
(b) 118° (e) 149°
(c) 35.3° or 144.7° (f) 110.9°

**7** 0.5530

**8** (a) 69° or 111° (c) 54°
(b) 65° (d) 113°

## Exercise 7 (p. 93)

**1** 0.6630; 0.9485

**2** Each is $\dfrac{\sqrt{3} - 1}{2\sqrt{2}}$ {note that $\sin\theta = \cos(90° - \theta)$}

**4** 0.8545
**5** 0.8945; $-2$
**6** $2 + \sqrt{3}$
**7** 3.0777; 0.5407
**9** (1) 0.5592 (2) 0.4848
**10** (a) 2.4751 (b) 0.8098

## Exercise 8 (p. 96)

**1** 0.96, 0.28, 3.428
**2** 0.4838, 0.8752, 0.5528
**4** 0.9917, $-0.1288$
**5** (1) 0.9511 (2) 0.3090
**6** 0.5
**8** 0.5; 0.8660
**9** 0.6001 approx.
**12** 0.268 approx.

## Exercise 9 (p. 99)

**1** $\frac{1}{2}(\sin 4\theta + \sin 2\theta)$
**2** $\frac{1}{2}(\sin 80° - \sin 10°)$
**3** $\frac{1}{2}(\cos 80° + \cos 20°)$
**4** $\frac{1}{2}(\sin 8\theta - \sin 2\theta)$
**5** $\frac{1}{2}\{\cos 3(C + D) + \cos (C - D)\}$
**6** $\frac{1}{2}(1 - \sin 30°) = \frac{1}{4}$
**7** $\cos 2A - \cos 4A$
**8** $\frac{1}{2}(\sin 6C - \sin 10D)$
**9** $2 \sin 3A \cos A$
**10** $2 \cos 3A \sin 2A$
**11** $2 \sin 3\theta \sin (-\theta)$
**12** $2 \sin 3A \sin 2A$
**13** $2 \cos 41° \cos 6°$
**14** $2 \cos 36° \sin 13°$
**15** $\cot 15°$
**16** $\tan \frac{\alpha + \beta}{2}$

## Exercise 10 (p. 102)

**1** $b = 15.8$; $c = 14.7$
**2** $a = 20.3$; $c = 30.4$
**3** $a = 7.18$; $c = 6.50$
**4** $c = 7.88$, $b = 5.59$
**5** $c = 17.3$; $a = 23.1$

## Exercise 11 (p. 104)

**1** $A = 28.95°$, $B = 46.59°$, $C = 104.48°$
**2** $A = 40.12°$, $B = 57.9°$, $C = 81.98°$
**3** $A = 62.18°$, $B = 44.43°$, $C = 73.38°$
**4** $A = 28.9°$, $B = 32°$, $C = 119.1°$
**5** 106.2°
6 43.85°

## Exercise 12 (p. 109)

**1** 114.4°
**2** 29.87°
**3** 45.45°
**4** $A = 22.3°$, $B = 31.47°$, $C = 126.23°$

**5** 65°; 52.33°; 62.67° (all approx.)
**6** 38.87°

## Exercise 13 (p. 113)

**1** A = 25.5°; C = 46.5°
**2** A = 64.32°; B = 78.28°
**3** B = 99.77°; C = 16.57°
**4** 83.42°; 36.58°
**5** 87.03°; 63.73°
**6** 65.08°; 42.68°

## Exercise 14 (p. 117)

**1** A = 29.4°; B = 41.73°; C = 108.87°
**2** A = 51.32°; B = 59.17°; C = 69.52°
**3** A = 43.32°; B = 35.18°; C = 100.3°
**4** A = 21.77°; B = 45.45°; C = 112.78°
**5** A = 35.38°; B = 45.67°; C = 98.95°

## Exercise 15 (p. 119)

**1** a = 166.5; B = 81.4°; C = 38°
**2** c = 172; A = 32.7°; B = 66.33°
**3** b = 65.25°; A = 33.43°; C = 81.42°
**4** c = 286.4°; A = 65.3°; B = 36.7°
**5** b = 136.6°; A = 58.63°; C = 90.92°

## Exercise 16 (p. 120)

**1** b = 145.2, c = 60.2, B = 81.47°
**2** a = 312, c = 213, C = 42.68°
**3** b = 151.4, c = 215, B = 42.05°
**4** a = 152.7, b = 83.4, A = 97.68°
**5** a = 8.27, c = 16.59, C = 110.9°

## Exercise 17 (p. 122)

**1** Two solutions: a = 4.96 or 58;
A = 126.07° or 3.93°
C = 28.93° or 151.07°
**2** Two solutions: a = 21.44 or 109.2
A = 11.32° or 88.68°
C = 128.68° or 51.32°

**3** One solution: b = 87.08, A = 61.3°, B = 52.7°
**4** Two solutions: b = 143 or 15.34
A = 35° or 145°
B = 115.55° or 5.55°

## Exercise 18 (p. 124)

**1** 19.05 $m^2$
**2** 72.36 $km^2$
**3** 39.42°
**4** 2537 $cm^2$
**5** 485 $cm^2$
**6** 64.8 $mm^2$
**7** 361.3 $mm^2$
**8** 24.17 $m^2$
**9** 0.503 Mg
**10** 239.6 $cm^2$
**11** 10 cm

## Exercise 19 (p. 125)

**1** 5.94 km
**2** A = 88.07°, B = 59.93°, C = 52°
**3** B = 45.2°, C = 59.57°, a = 726
**4** C = 56.1°
**5** 16.35 m, 13.62 m
**6** 41°
**7** Two triangles: B = 113.17° or 66.83°
C = 16.83° or 63.17°
c = 9.45 or 29.1
**8** 267 m approx.
**9** 6.08 m, 5.71 m
**10** 3.09 mm
**11** 7.98 cm, P = 26.33°, a = 29.93°
**12** 4.5 cm, 6 cm; 11 $cm^2$
**13** $4\frac{1}{2}$ h
**15** 0.3052 $m^2$
**16** 49.47°; 58.75°

## Exercise 20 (p. 138)

**1** 15.2 m
**2** 546 m
**3** 276 m
**4** 193 m approx.
**5** 889 m approx.
**6** 1.26 km
**7** 3700 m
**8** 11 990 m
**9** 2.88 km approx.
**10** 2.170 km
**11** 500 m approx.
**12** 3.64 km; 45° W. of N.; 5.15 km
**13** 73 m; 51 m
**14** 1246 m approx.
**15** 189 m approx.
**16** 63.7 m approx.
**17** 1970 m and 7280 m approx.

## Exercise 21 (p. 145)

**1** 60°, 15°, 270°, 120°, 135°
**2** (a) 0.5878 (b) 0.9239 (c) 0.3090 (d) 0.3827 (e) 0.9659
**3** (a) 4.75 (b) 2.545
**4** (a) 13.4° (b) 89.38°
**5** (a) $\frac{\pi}{12}$ (b) $\frac{2\pi}{5}$ (c) $\frac{11\pi}{30}$ (d) $\frac{7\pi}{12}$
**6** (1) 5.842 cm (2) 17.5 m
**7** $\frac{11}{18}$ radians; 35°
**8** 1.57 approx.
**9** $\frac{\pi}{4}$; $\frac{\pi}{3}$; $\frac{5\pi}{12}$

## Exercise 22 (p. 161)

**1** (a) −9.9744; −0.2250; 4.3315
(b) −0.3619; −0.9322; 0.3882
(c) −0.7030; 0.7112; −0.9884
(d) −0.2901; 0.9570; −0.3032
**2** (a) −0.7771 (b) 0.7431 (c) −0.6691 (d) −0.2419
**3** (a) −1.0576 (b) 2 (c) −1.2349 (d) −1.7434
**4** (a) −0.8387 (b) 0.7431 (c) 1.2799 (d) 0.5878

## Exercise 23 (p. 170)

**1** (1) 63°, 117° (2) 65.3°, 294.7° (3) 19.3°, 199.3° (4) 65.1°, 294.9°
**2** (1) 20.7°, 159.3° (2) 18.43°, 71.57°
**3** (1) 0°, 180°, 80.53°, 279.58°
(2) 43.87°, 136.13°
(3) 45°, 135°, 225°, 315°
(4) 30°, 150°, 210°, 330°
**4** (1) 26.57°, 45°, 206.57°, 225°
(2) 60°, 270°, 300°
(3) 60°, 300°
(4) 0°, 120°, 180°, 240°

**5** (1) $2n\pi \pm \cos^{-1} 70.8°$
(2) $n\pi + (-1)^n \sin^{-1} 19.7°$

(3) $n\pi$ or $n\pi + (-1)^n \frac{\pi}{6}$

(4) $n\pi + \frac{\pi}{12}$ or $n\pi + \frac{5\pi}{12}$

**6** (1) 13.03°
(2) 53.13°
(3) 6.48°
(4) 36.87°

TEACH YOURSELF BOOKS

# ALGEBRA

## P. Abbott and M.E. Wardle

This well-known introduction to algebra has been fully updated by Michael Wardle, Senior Lecturer in Mathematics Education at the University of Warwick.

Covering the ground from the very beginning the text explains algebra's various elements such as equations, factors and indices, continuing on to simple progressions and permutations. The course is carefully graded and only a knowledge of elementary arithmetic is assumed.

'Granted that it is possible to learn the subject from a book, then this volume will serve the purpose excellently . . . A chapter worthy of special mention is that dealing with determination of graph laws, a topic too often neglected but one of great value and frequent practical use.

Regarded as a textbook, this is probably the best value for money on the market.'

*Higher Education Journal*

TEACH YOURSELF BOOKS

# CALCULUS

## P. Abbott and M.E. Wardle

This respected text has been revised by Michael Wardle, Senior Lecturer in Mathematics Education at the University of Warwick.

Ideal for beginners, this book provides a series of lessons which introduce the basic concepts of differentiation and integration. Each chapter includes many clearly worked examples, diagrams and exercises.

'A student with a sound knowledge of algebra and trigonometry should be able to learn calculus by studying this excellent book.'

*Scottish Education Journal*